Do Joint Fighter Programs Save Money?

Technical Appendixes on Methodology

Mark A. Lorell, Michael Kennedy, Robert S. Leonard, Ken Munson,
Shmuel Abramzon, David L. An, Robert A. Guffey

RAND Project AIR FORCE

Prepared for the United States Air Force
Approved for public release; distribution unlimited

RAND
CORPORATION

The research described in this report was sponsored by the United States Air Force under Contract FA7014-06-C-0001. Further information may be obtained from the Strategic Planning Division, Directorate of Plans, Hq USAF.

Library of Congress Cataloging-in-Publication Data is available for this publication.

ISBN: 978-0-8330-7932-9

The RAND Corporation is a nonprofit institution that helps improve policy and decisionmaking through research and analysis. RAND's publications do not necessarily reflect the opinions of its research clients and sponsors.

Support RAND—make a tax-deductible charitable contribution at www.rand.org/giving/contribute.html

RAND® is a registered trademark.

RAND OFFICES
SANTA MONICA, CA • WASHINGTON, DC
PITTSBURGH, PA • NEW ORLEANS, LA • JACKSON, MS • BOSTON, MA
DOHA, QA • CAMBRIDGE, UK • BRUSSELS, BE
www.rand.org

Preface

Joint aircraft programs, in which two or more services participate in the development, procurement, and sustainment of a common aircraft design, are thought to save life cycle cost (LCC) by eliminating duplicate efforts and realizing economies of scale.

In theory, joint programs have more potential to save costs than multiple comparable single-service programs by sharing total research, development, test, and evaluation (RDT&E) expenditures on a common design, and achieving economies of scale in production and operations and support (O&S). But the need to accommodate different service requirements in a single design or common design family may lead to greater program complexity, increased technical risk, and common functionality or increased weight in excess of that needed for some service variants, potentially leading to higher overall cost despite these efficiencies. The fundamental question we seek to answer is this: On average, are the theoretical savings that should accrue from joint aircraft programs sufficient to offset the additional costs arising from greater complexity? In short, do joint fighter and other aircraft programs cost less overall throughout their entire life cycle than an equivalent set of specialized single-service systems?

RAND Project AIR FORCE analyzed the costs and savings of joint tactical aviation acquisition programs to determine whether a joint approach achieves the anticipated cost savings. The study team examined whether historical joint aircraft programs, and the Joint Strike Fighter (JSF) in particular, have saved LCC compared with comparable notional single-service programs. The team also examined the implications of joint fighter programs for the health of the industrial base and for operational and strategic risk. The major study findings are documented in a separate report.[1]

This report provides a series of appendixes that detail the methodology behind the study findings. It is intended for analysts who wish to examine or replicate the RAND analysis. The research reported here was sponsored by Gen Donald Hoffman, former Commander of Air Force Materiel Command, and was conducted within the Resource Management Program of RAND Project AIR FORCE as part of a project titled "Cost/Benefit Analysis of Joint Tactical Aviation Acquisition Programs."

---

[1] Mark A. Lorell, Michael Kennedy, Robert S. Leonard, Ken Munson, Shmuel Abramzon, David L. An, and Robert A. Guffey, *Do Joint Fighter Programs Save Money?* Santa Monica, Calif.: RAND Corporation, MG-1225-AF, 2013.

RAND Project AIR FORCE

RAND Project AIR FORCE (PAF), a division of the RAND Corporation, is the U.S. Air Force's federally funded research and development center for studies and analyses. PAF provides the Air Force with independent analyses of policy alternatives affecting the development, employment, combat readiness, and support of current and future air, space, and cyber forces. Research is conducted in four programs: Force Modernization and Employment; Manpower, Personnel, and Training; Resource Management; and Strategy and Doctrine.

Additional information about PAF is available on our website:
http://www.rand.org/paf/

Contents

Figures

Tables

Summary

The U.S. Department of Defense (DoD) has launched or attempted to launch numerous joint fighter and other joint aircraft programs in the past 50 years. These programs were intended to save life cycle cost (LCC) by eliminating duplicate research, development, test, and evaluation (RDT&E) efforts and achieving economies of scale in procurement and operations and support (O&S). Thus far, there have been no comprehensive assessments of empirical data to verify that joint aircraft programs have, indeed, saved LCC compared with an equivalent set of specialized single-service aircraft systems. Although it is acknowledged that the need to integrate multiple-service requirements in a single design or common design family can increase programmatic and technical complexity and lead to performance and cost penalties with greatest common denominator designs, analysts are unsure how large these factors are and the degree to which they reduce or even negate the efficiency savings from having only one program. Analysts are also unsure of the degree to which the effects of these factors are accurately estimated.

In the absence of direct cost comparisons in which multiple similar single-service programs were developed in parallel with an equivalent joint program, RAND Project AIR FORCE (PAF) sought to answer the question of which approach costs less by comparing the cost growth of joint versus single-service aircraft programs. If cost growth tends to differ and be higher for joint aircraft programs, this would suggest that the difficulties of joint, common programs are typically underestimated. The degree of underestimation, if any, can be used to estimate whether total costs become higher or lower compared to single service programs.

The ultimate question we seek to answer with our full methodology is whether, in the end, the actual realized cost benefits of joint aircraft programs offset and exceed any increased costs due to greater complexity, resulting in a force of aircraft with lower LCC than an equivalent force of specialized single-service aircraft.

PAF sought to answer this question by assessing historical joint aircraft program outcomes and cost data from the early 1960s through today's Joint Strike Fighter (JSF). Among the major findings, PAF found that historical joint aircraft programs have experienced rates of acquisition cost growth so much higher than single-service programs that they have not saved overall LCC despite any efficiencies from common efforts. Researchers also found that, nine years after Milestone B (MS B), the JSF program is not on the path to achieving the LCC savings expected at MS B compared with three comparable notional single-service fighter programs. These findings are presented in a separate report.[1]

This report provides a series of appendixes that describe the underlying methodology. The first four detail the methods used to estimate whether historical joint aircraft programs have saved LCC over the likely LCC had single-service programs been pursued instead, as reported in Chapter Two of the main report, particularly Sections 1 and 2:

- Appendix A describes how researchers calculated the theoretical maximum savings a joint aircraft program could achieve in the acquisition phase (i.e., in RDT&E and procurement), assuming 100 percent commonality between the joint aircraft design variants procured by each participating service, and two 100 percent common single-service aircraft developed and produced entirely separately by each service. Using algebraic formulae and reasonable assumptions, researchers found that an "ideal" joint aircraft program can save a maximum of 20 percent of acquisition costs compared with two single-service programs.
- Appendix B shows how researchers compared acquisition cost-growth rates for four historical joint aircraft programs and four historical single-service aircraft programs.[2] This analysis leads to the finding that, nine years past MS B,[3] the joint programs experienced an additional 41 percent cost growth on average compared with the single-service programs. This amount of excess program cost growth experienced by joint programs would eliminate any joint savings realized during the acquisition phase, even in an ideal joint program.
- Appendix C shows how researchers calculated the maximum joint O&S savings that can be achieved in an ideal joint program. Reviewing empirical data, researchers found that an ideal two-service joint fighter program (with each program having an equal number and mix of the same type of aircraft and 100 percent common O&S activities) could save a maximum of 2.9 percent in O&S costs.
- Appendix D completes the analysis of historical joint aircraft programs by analyzing the O&S cost savings that joint aircraft programs would need to achieve to offset the greater average cost growth experienced by joint programs during the acquisition phase in order to achieve overall LCC savings compared to the LCC of equivalent single-service programs. Researchers found that a typical

---

[1] Mark A. Lorell, Michael Kennedy, Robert S. Leonard, Ken Munson, Shmuel Abramzon, David L. An, and Robert A. Guffey, *Do Joint Fighter Programs Save Money?* Santa Monica, Calif.: RAND Corporation, MG-1225-AF, 2013.

[2] Our analysis used the RAND Selected Acquisition Report (SAR) data base for consistent comparisons of cost estimates across programs. The eight aircraft programs examined represent the entire available body of Major Defense Acquisition (MDAP) aircraft programs from the mid 1980s through the present that have full cost data reported in the annual SARs out to at least nine years past MS B. However, classified and other special aircraft programs developed outside the normal DoD acquisition process, such as the B-2, are excluded because they have no SAR reported acquisition cost data available.

[3] Nine years after MS B was the longest period after MS B for which we had complete and comparable SAR data for all eight aircraft at the time the analysis was conducted in late 2011.

representative joint aircraft program would need to save more than 10.3 percent in O&S costs to offset the average joint acquisition cost growth premium described in Appendix B (even when assuming the theoretical maximum acquisition cost savings for an ideal joint program described in Appendix A). This showed that joint O&S savings, even under ideal conditions, are not sufficient to offset the average joint acquisition cost growth premium.

Using the methodology and findings reported in Appendixes A through D, we concluded that recent historical joint aircraft programs have not saved overall LCC compared with comparable single-service aircraft programs.[4]

The last two appendixes describe two separate methods used to compare JSF savings and costs with three notional comparable single-service fighters (which represent the "path not taken") and support the analysis reported in Chapter Three of the main report, particularly Sections 1 and 2:

- Appendix E describes the methodology for assessing the JSF presented in the report. Researchers made a set of plausible, conservative assumptions to calculate the LCC of three separate, notional single-service fighter aircraft programs before and after nine years of cost growth (assuming F-22 cost-growth rates for the single-service programs[5]). These costs were compared with the actual JSF cost estimates at MS B and nine years later (the most-recent SAR data available at the time of the study). This analysis resulted in the conclusion that nine years after MS B, JSF is not on course to providing the savings that were expected at MS B compared with three separate notional single-service programs.
- Appendix F presents an alternative method of comparing JSF procurement costs with those of three notional single-service fighter programs, using weight-cost relationship curves based on F-22 program data and historical aircraft program cost data for the procurement phase. Researchers arrived at a very similar cost relationship as the methodology described in Appendix E.

The findings based on the methodology reported in Appendixes E and F led us to conclude that the JSF F-35 program is not on the path to providing the joint savings anticipated at MS B.

The main report provides further analysis of the qualitative factors that drive these cost results. Together with the quantitative analysis detailed in this report, they support PAF's overall recommendation that, unless the participating services have identical,

---

[4] Our data show that the average joint cost-growth premium varies between 30 and 44 percent from five years out beyond MS B through nine years beyond MS B. We used this variation to develop an uncertainty band for the comparison of joint versus single-service acquisition costs and LCC. In both cases, the findings at year nine fall within the uncertainty band. For this reason, we conclude that, on average, historical joint programs have not saved overall costs for either the acquisition program or the overall LCC including RDT&E, procurement, and O&S costs.

[5] Different time frames for calculating F-22 cost growth rates as well as considerable sensitivity analysis were conducted to increase confidence in our analysis.

stable requirements, DoD should avoid future joint fighter and other complex joint aircraft programs.

Acknowledgments

We are grateful for the broad expertise, advice, constructive criticism, and assistance provided by numerous Air Force officers, as well as colleagues at RAND, without whom this study could not have been completed. Gen Donald Hoffman, Commander of Air Force Materiel Command (AFMC),[1] initiated this research and provided crucial guidance throughout. We would also like to acknowledge Lt Gen (ret.) George K. Muellner, former Principal Deputy, Office of the Assistant Secretary of the Air Force for Acquisition, and former Director and Program Executive Officer, Joint Advanced Strike Technology Program, for his invaluable insights into the early genesis of the JSF program. William Suit, Headquarters AFMC History Office, searched through vast amounts of primary source materials and provided extensive archival documentation that proved critical for our evaluation of historical joint fighter programs. Extensive information on the F-111 program derived from her careful study of the documentary record was graciously shared with us by Molly J. Waters, Operations Research Analyst, Headquarters AFMC. Former Lt Col Michael McGee, Operational Requirements Lead, JSF, Directorate of Operational Requirements, Headquarters U.S. Air Force, during the early phases of the JSF program, contributed useful information on the early requirements-reconciliation process.

Many RAND colleagues contributed crucial analysis, insights, critiques, and advice in the course of this research. In this regard, we are especially grateful for the contributions of Fred Timson, James S. Chow, Obaid Younossi, John C. Graser, Donald Stevens, Daniel M. Romano, Natalie W. Crawford, Paul DeLuca, and many others. Eric Peltz, Associate Director of the RAND National Security Research Division and Director of the RAND Supply Chain Policy Center, provided a comprehensive critique of an earlier version of this document that led to significant improvements.

The constant support, substantial substantive input, and critical guidance provided by Andrew R. Hoehn, Senior Vice President for Research and Analysis for RAND and former Director of Project AIR FORCE; Lara Schmidt, Associate Director of RAND Project AIR FORCE; and Laura H. Baldwin, Director of the Resource Management Program of RAND Project AIR FORCE, are greatly appreciated. We also owe a special debt of gratitude to our project monitor, Ross Jackson, Deputy Division Chief, Studies and Analyses Division, Headquarters AFMC, for his considerable assistance and encouragement.

---

[1] All ranks, positions, titles, and offices are as of the time of the research.

Abbreviations

ACC	Air Combat Command
AETC	Air Education and Training Command
AFMC	Air Force Materiel Command
AFRC	Air Force Reserve Command
AFTOC	Air Force Total Ownership Cost
ANG	Air National Guard
APUC	average procurement unit cost
BY	base year
CG	cost growth
CIC	cost improvement curve
CLS	contractor logistics support
CTOL	conventional takeoff and landing
CV	aircraft carrier hull-type designation
DLR	depot-level reparable
DoD	Department of Defense
DON	Department of the Navy
EW	empty weight
FH	flying hour
FY	fiscal year
IOC	initial operational capability
JPATS	Joint Primary Aircraft Training System
JPO	joint program office
JSF	Joint Strike Fighter
JSTARS	Joint Surveillance Target Attack Radar System
LCC	life cycle cost
LIMS-EV	Logistics Installations and Mission Support–Enterprise View
LO	low observable
LRIP	Low Rate Initial Production
MDAP	major defense acquisition program
MILCON	military construction
MS B	Milestone B
O&S	operations and support
OSD	Office of the Secretary of Defense
PACAF	Pacific Air Forces

PAF	Project AIR FORCE
RDT&E	research, development, test, and evaluation
R&D	research and development
SAR	Selected Acquisition Report
STOVL	short takeoff and vertical landing
TAI	total aircraft inventory
USAFE	U.S. Air Forces in Europe

Appendix A: Calculation of Theoretical Maximum Joint Aircraft Acquisition Program Savings

As the first step in determining whether joint aircraft programs on average save overall life cycle cost (LCC) compared with a set of comparable specialized single-service aircraft programs, we calculated the theoretical maximum amount of cost savings for an ideal joint program compared with comparable single-service programs. The methodology we used is explained below. The analysis shown in Appendix A (as well as in Appendixes B, C, and D) supports the findings reported in Chapter Two of the main report.[1]

This appendix addresses the question: What is the highest level of acquisition cost savings that can be realized by joint, versus single-service, programs? In this appendix, we calculate the cost difference between two notional programs. The first is a program in which two services jointly develop a single identical aircraft and procure it in equal quantities. The second includes two identical single-service programs—that is, programs in which each service independently develops and procures the same number of an identical aircraft, with the sum of the two services' aircraft production equal to the total production of the joint program. This calculation leads to the "maximum" level of savings, because there is no penalty whatsoever to combining the two single-service programs. Since the aircraft are identical, the second research and development (R&D) program is completely redundant, and the two services' aircraft can be produced on a single production line, taking full advantage of economies of scale. In real programs, there would be less commonality of course, so the savings would be less.

Let R be the dollar cost of the R&D program that is needed to develop the aircraft. The joint program will incur this R&D cost, and the two single-service programs will incur a cost $2 \times R$, so the joint R&D savings are 50 percent. This is the maximum that can be achieved; if the two services have different variants, there will be unique content in the R&D required for each, so the cost of the joint program would be more than half the cost of two identical single-service programs.[2]

---

[1] See Mark A. Lorell, Michael Kennedy, Robert S. Leonard, Ken Munson, Shmuel Abramzon, David L. An, and Robert A. Guffey, *Do Joint Fighter Programs Save Money?* MG-1225-AF, 2013.

[2] We note that the increase in research, development, test, and evaluation (RDT&E) cost due to unique content can result from two distinct phenomena. The first is simply the cost of developing the unique components. The second is any additional complexity or system integration cost that results from combining the unique components with the common ones. We note that some other costs to the Department of Defense (DoD), which are not generally captured in program costs, could also be lowered. These would

To determine production cost savings, we must specify a learning, or cost-quantity improvement rate. For this example, we specify an 87 percent cost-quantity improvement curve. This is a representative rate for military aircraft programs.[3] Define $CC(n)$ as the cumulative cost of producing n units of the aircraft. Then, an 87 percent cost-quantity improvement curve is represented by the equation[4]

$$CC(n) = A(n)^{\beta}.$$

Here, A is the cost of the first unit, and β is a coefficient defined by

$$\beta = \left[\frac{\ln(\lambda)}{\ln(2)}\right] + 1,$$

where λ is the learning rate expressed as a percentage (i.e., 0.87). In this case $\beta = 0.8$. Note that the above expression is equivalent to

$$\lambda = (2)^{(\beta-1)}.$$

Say that N units of the identical aircraft would be produced in each of the two service's individual programs, so that there would be $2N$ produced in a joint program. The cost of two single-service programs would be

$$2A(N)^{\beta},$$

and the cost of a joint program would be

$$A(2N)^{\beta}.$$

Thus, the savings of a joint program are

$$2A(N)^{\beta} - A(2N)^{\beta},$$

and the percentage savings is

---

primarily be the costs of originally setting up the program and running the source-selection process, as well as the general costs of operating two program offices rather than one (presumably larger) joint office.

[3] There is no standard "right" number for the learning rate of a fighter-attack program. We chose 87 percent as a representative value. It is the average of two recent fighter-attack programs: The Milestone B (MS B) learning rate of the F-22 was 90 percent, and that of the F/A-18E/F was 84 percent. But, again, it was chosen as a representative value, not based on any statistical methodology.

[4] We chose this functional form for this appendix because it represents a constant learning rate that relates cumulative cost to cumulative production. In our professional judgment, it was the most useful functional form to illustrate the principles involved. One could choose other functional forms. In particular, some analysis uses a functional form that relates unit cost, rather than cumulative average cost, to cumulative production. Others use a "broken" learning relation in which the learning rate changes as cumulative production occurs. In addition, one could add a rate effect to the analysis. That is, unit cost in any period would be related to both cumulative production and the production level in that period. There is no professional consensus as to what is the "right" functional form.

$$\frac{\left[2A(N)^\beta - A(2N)^\beta\right]}{\left[2A(N)^\beta\right]}, \text{ or}$$

$$1-(2)^{\beta-1},$$

which is simply one minus the learning, or cost-improvement, rate of the program (λ) expressed as a percentage, or 0.13 if the learning rate is 0.87.

The total acquisition savings are the sum of the R&D and production savings, which is equal to

$$R+\left[2A(N)^\beta - A(2N)^\beta\right].$$

The percentage savings will be a weighted average of the R&D and production percentage savings, with the weights representing the share of R&D in the single-service acquisition programs and the share of production in the single-service acquisition programs, respectively. Let these shares be defined by S_R and S_P. They are the same for both of the single-service programs, since the programs are assumed to be identical. Then, the total percentage acquisition savings from a joint program will therefore be

$$(0.5)S_R+[1-\lambda]S_P.$$

Typical RDT&E and procurement shares for a fighter program are 0.2 and 0.8, respectively.[5] Given the 50 percent R&D saving of the joint program, and the 13 percent procurement saving (at an 87 percent learning rate), the total saving would be 20 percent. Differing assumptions about the R&D and procurement shares in the single-service programs, and about the production learning rate, would give different answers, of course.

---

[5] Appendix C discusses the issue of representative RDT&E and procurement shares. As is noted in Appendix C, the 20/80 split between RDT&E and procurement was deemed typical for an aircraft acquisition program with substantial procurement quantities. This split is determined by the complexity of the development effort and the number of production aircraft. For a tactical fighter, development can be as low as 12 percent and as high as 60 percent of acquisition costs. The low end of the range is for a derivative fighter design, such as the F/A-18E/F; the high end could occur if the aircraft is highly complex and, more important, if the quantity of aircraft to be acquired is drastically reduced. The F-22 makes a good example. We chose 20 percent as representative value, assuming that a substantial production run would be included in the program.

Appendix B: Calculation of the Joint Acquisition Cost-Growth Premium from Historical Aircraft Programs

As illustrated in Appendix A for acquisition costs, and in Appendix C for operations and support (O&S) costs, the two potential cost-savings mechanisms of one joint program over multiple single-service programs come from eliminating one or more R&D efforts and economies of scale affecting procurement and sustainment costs. Those appendixes calculate the maximum plausible saving in their applicable element(s) of LCC. In this appendix, we first discuss potentially offsetting effects and the reasons that joint program savings might be overestimated, and then we calculate the observed excess cost growth in joint programs compared with single-service programs.

Why Joint Program LCC Savings Are Likely to Be Less Than Their Theoretical Maximums

The following characteristics of joint programs may counteract the cost savings potential of joint programs compared with single-service programs:

- more costly R&D efforts because of the complexity and risk of combining a broader array of requirements into a single common design family
- more costly procurement and sustainment when the platforms and parts themselves become more expensive because of design to the greatest common denominator.

More specifically, achieving the greatest common denominator in all joint program units may require adding capabilities and weight to the common platform above those that would be needed in each individual platform if they were designed separately. There is a cost throughout the life cycle for excess functionality or weight of the common platform compared with the optimal design for each functional variant.

These potential countervailing joint program effects can be represented in either a less-optimistic MS B estimate in the form of assumed savings lower than those shown in Appendixes A and C or higher cost growth from the MS B estimate as the effects take hold over the life cycle of the joint program. Review of program histories indicates that joint programs are evaluated and selected on the basis of their potential theoretical maximum benefits under ideal conditions; thus, these countervailing effects are more likely to emerge via higher cost growth. This appendix illustrates how we tested for such cost growth and calculated its scale.

5

In addition, multiple single-service programs with separate platforms may also share some common components with the other related single-service program(s) or with other nonaffiliated programs (such as F-22 components in the Air Force version of the F-35), which provide some of the potential benefits of one joint program without the potential negative cost effects.[1] This positive effect decreases the costs of multiple single-service programs in all LCC and therefore reduces any real savings that a single joint program might have over multiple single-service programs. It is likely that these savings would be accounted for at the MS B of any real-world single-service program to "sell" the program, giving an additional reason that the theoretical maximum saving calculated in Appendixes A and C are unlikely to fully materialize.

The Challenge of Estimating Real Differences in Costs Between Joint and Single-Service Programs

Unfortunately, there have been no cases where joint common and similar single-service programs have proceeded simultaneously, thus allowing direct measurement of countervailing effects. Nor have full program costs for these two alternative directions been developed sufficiently well to permit meaningful comparisons of actual cost data.

In the absence of these direct comparisons, the RAND team developed a way to compare cost growth to estimate the effects of the positive and negative forces at play on joint aircraft programs, or at least to estimate the degree to which the positive forces tend to be overestimated and the negative forces underestimated. This stems from the premise that if it is assumed that the beneficial effects of joint aircraft programs occur with little of the countervailing effects (e.g., a very high degree of commonality is maintained with no excessive weight or complexity resulting from differing service requirements) but the reality is different, then the joint program is likely to experience more cost growth than comparable single-service programs. We sought to use actual program data to assess whether historical joint aircraft programs experienced greater cost growth than single-service programs. The higher cost growth we found for past joint aircraft programs compared with single-service programs indicates that DoD has tended to underestimate the additional joint R&D complexity and risk, overestimate the commonality, or

---

[1] Indeed, in the early 1990s, before the JSF program, the Air Force and the Navy were planning to develop separate single-service fighters but were discussing using the F119 engine planned for the F-22 to power both single-service aircraft. If this had been carried out, the Air Force and the Navy single-service fighters could have enjoyed a cost savings on their engines from economies of scale by using the same engine on their new fighter as well as on the F-22. This could have resulted in a significant savings, since, as a general rule of thumb, fighter production costs are typically split fairly evenly among airframe, engine, and avionics.

underestimate the excess functionality/weight required when developing joint aircraft program cost estimates.

In effect, the patterns show that the reality of commonality tends to give more weight to the challenges than the assumed benefits. Instead of the joint R&D effort being similar to a single-service R&D effort, it becomes considerably more complex, difficult, and costly than expected, and the cost of a platform becomes higher than expected, affecting procurement and likely O&S cost growth as well.

Analytical Data Source and Definitions

The RAND Selected Acquisition Report (SAR) database containing some 300 major defense acquisition programs (MDAPs) initiated from the late 1960s through 2010 was the source of information for the acquisition programs and their cost data that we used to compare cost growth from MS B in joint versus single-service aircraft programs.[2]

Originally, the RAND team had hoped to compare a significant number of joint fighter MDAPs to single-service fighter MDAPs. A quick survey of the historical record showed that this was not possible. We determined that 11 major joint fighter programs had been proposed between 1960 and 1995.[3] However, of these, only four went beyond the initial proposal stage (the General Dynamics F-111, the McDonnell F-4, the Ling Tempco Vought A-7, and the JSF). Of these, two (the F-4 and the A-7) did not meet our definition of a fully joint program, in that they were designed and developed entirely by one service, then only procured jointly after single-service development. Of the remaining two, only one (the JSF) proceeded beyond the initial development stage. Thus, to find a reasonable sample size, we were forced to expand our search beyond fighters and include all military aircraft MDAPs. We also had to impose criteria for time period and time phasing of the programs to ensure sufficient, reliable, and comparable cost data.

The following criteria were used to select the programs for comparison:

- It is a fixed-wing aircraft acquisition program.
- It requires substantial development effort.
- It is not a modification program but rather a program for the acquisition of an entirely new aircraft.
- It was initiated (MS B) in the mid-1980s or more recently.

---

[2] The data cutoff point for this research was November 2011. Therefore, the most-recent SARs available at that time were the December 2010 SARs.

[3] See Table 2.1 in the main report, Lorell et al., 2013.

- It has a MS B cost baseline estimate.[4]
- It is at least nine years past its MS B as of its December 2010 SAR.[5]

The primary factor distinguishing joint from single-service programs was multiservice involvement from at least program initiation at MS B with substantial development funding by each participant and planned continued multiservice participation throughout the life cycle of the weapon system.

Few aircraft programs met all criteria for either group. We found only four joint and four single-service aircraft MDAPs that met all the criteria. This is because many programs were either modifications to existing aircraft, were versions of existing civilian or military aircraft and therefore did not have a MS B (program initiation at MS C), or were initiated before the mid-1980s or after 2001. To increase the number of programs in the data set, we considered expanding it to older programs but found that doing so would add only to the single-service data set; there were no joint aircraft programs meeting our remaining criteria that were begun before fiscal year (FY) 1985 for which we have MS B baseline SAR estimates. Adding older programs also ran the risk of including far less complex efforts than typical of a modern MDAP, so the relevance of the older programs data is less certain to future programs.

The eight programs—four single-service and four joint—are shown in Table B.1. Each set of four programs has a roughly comparable mix of technological complexity and challenge regarding the basic aircraft system development. All eight programs had MS B dates between FY 1985 and FY 2002, inclusive. The similar mix of technology challenge and MS B dates in the two data sets likely reduces potential bias that these factors might have on cost growth.

Of the joint program data set, both the F-35 and the V-22 involved three variants at their MS B.[6] Joint Surveillance Target Attack Radar System (JSTARS) is the only

---

[4] Traditionally, only larger (in dollar terms) aircraft programs with substantial development efforts begin SAR reporting at or before their MS B. In these programs, an official MS B baseline is prepared and reported in a SAR. Acquisition programs for such aircraft as the Predator and Reaper remotely piloted vehicles were not large enough to report via SARs at the beginning of their major development efforts and therefore could not be considered for inclusion in this data set.

[5] Extending the range beyond nine years after MS B was not possible because of insufficient available data for all eight programs at the time the research was conducted.

[6] In the V-22 program, all four services intended to acquire a variant of the platform at the time of the program's MS B. The Marine Corps desired the basic medium-lift variant, the Navy a combat search and rescue variant, the Air Force a special operations variant, and the Army planned to acquire the Marine Corps variant. All services contributed substantial funds to both the program's development and procurement funding. Shortly after reaching MS B, the Navy and Army scrapped their plans to acquire V-22s, leaving the Marine Corps and Air Force variants that are now operational.

program in which the aircraft segment of the overall system is operated by a single service.[7]

Table B.1
MDAP Programs for Joint Versus Single-Service Cost-Growth Comparison

Program Name	Single-Service or Joint	Technological Challenge	Participants	MS B Date
C-17	Single service	Medium	Air Force	December 1985
F/A-18E/F	Single service	Medium	Navy	July 1992
F/A-22	Single service	High	Air Force	August 1991
T-45 TS	Single service	Low	Navy	October 1984
JPATS T-6A/B	Joint	Low	Air Force and Navy	February 1996
F-35 (JSF)	Joint	High[a]	Air Force, Navy, and Marine Corps	October 2001
V-22 Osprey	Joint	High	Marine Corps, Air Force, and Army	May 1986
E-8A (JSTARS)	Joint	Medium	Air Force and Army	September 1985

[a] Two of the variants, the conventional takeoff and landing (CTOL) and CV (an aircraft carrier hull-type designation), could be characterized as Medium. Only the short takeoff and vertical landing (STOVL) variant is clearly High. The overall characterization of High is largely a function of jointness and the need to accommodate the STOVL variant into the program with maximum possible commonality with the two other variants.

For each program in the data set, cost growth for its development funding, procurement funding, and program total were calculated from the program's MS B to approximately nine years past that milestone.[8] At this point past MS B, SAR estimates are a mixture of nine years (or more if the programs had pre-MS B funding) historical funding plus estimates of all future funding. In aircraft programs, Low Rate Initial Production (LRIP) is typically under way at the nine-year point past MS B. Much, but not all, of the development has been funded, and therefore the RDT&E estimate is in large part historical actual appropriations and in smaller part planned future appropriations. Most production still lies ahead in the program, thus production estimates are only in

---

[7] In March 1984, 18 months before the JSTARS MS B, the joint program office planned to field three separate airborne radar platforms. This approach changed shortly thereafter with the determination that the requirements of both services could be provided on a single, more-complex platform. That platform acquires and transmits the required imagery to both Army and Air Force users. This program structure was determined to be the most efficient form of a joint aircraft program as only one variant was needed to fulfill all requirements of both services.

[8] Nine years after MS B was chosen as the end point because this was the furthest point beyond MS B for which we had reasonably complete sets of cost data for all eight aircraft programs at the time of the data cutoff point for this research, which was November 2011.

small part historical actual appropriations and in large part planned future appropriations. Because the future production estimates are based on some actual production experience from LRIP, it is likely that they are based on verifiable experience rather than pure estimation.

The calculations used for our comparisons are done in each program's base-year (BY) dollars as provided in the SARs. This removes the effects of inflation and changes in inflation, thus "real" buying power cost growth is measured. Cost-growth factors for procurement and total program costs are adjusted for quantity changes. The adjustment is made to the quantity of aircraft planned to be procured as of nine years past MS B or the MS B quantity, whichever is lower. The lower numbers are used because estimates for these aircraft exist both at the time of the MS B and at the nine-year point thereafter. If quantity is cut from MS B, then cost growth is measured on the planned number of units at nine years past MS B, which is fewer than the number at MS B. If quantity is increased from MS B, then cost growth is measured on the MS B quantity, excluding the units added to the program.

Table B.2 shows the units for which cost growth was calculated in each program. In all four single-service programs, the quantity planned at nine years past MS B is lower than that at MS B, with quantities cut to 49 percent, or less than half, on average.[9] In the joint programs, two are lower at nine years past MS B and two are higher; on average, the quantities on which cost growth is measured after nine years were 86 percent of the MS B estimate.

---

[9] The C-17 quantity is far below its MS B quantity, at just 19 percent. The program had production problems at the time, such that future production was put on hold pending a production process overhaul to reduce future unit costs.

Table B.2
MDAP Dataset Quantities at Nine Years Past MS B

Program Name	MS B Quantity	Quantity at Nine Years Past MS B	Quantity for Which Cost Growth Was Measured	Percentage of MS B Quantity on Which Cost Growth Was Measured
C-17	210	40	40	19
F/A-18E/F	1,000	548	548	55
F/A-22	648	315	315	49
T-45 TS	300	218	218	73
Single-service program average percentage of MS B quantity				**49**
JPATS T-6A/B	711	782	711	100
F-35 (JSF)	2,852	2,443	2,443	86
V-22 Osprey	913	523	523	57
E-8A (JSTARS)	10	19	10	100
Joint program average percentage of MS B quantity				**86**

This difference in the data sets could be significant, because the later production units in any program tend to experience more cost growth than do the earlier units. This occurs because MS B estimates tend to use a constant cost-quantity improvement curve over the life of the production run, meaning continuously lower unit costs throughout production. In reality, unit costs tend to somewhat flatten out over time because of design enhancements and lower than full production rates later in the production run; thus, later unit cost decreases are lower than estimated at MS B.

With such a large difference between the two data sets in the fraction of MS B quantity planned to be built at nine years past MS B, there is the potential for a bias toward more cost growth in the joint programs. This effect is likely minor, because only a small fraction of total planned units are built at that point in the typical military aircraft program, future enhancements are typically not yet part of the current program plan, and full rate production remains the plan for the balance of the program. To test this, we measured the cost growth in the first 49 percent of the units planned at MS B in the four joint programs. We found that doing so reduces the joint program cost-growth penalty by less than one-tenth, from 41.2 percent to 37.3 percent.[10]

Table B.3 shows acquisition cost growth in BY 2012 dollars and the associated percentages for each of the eight programs. The bottom data row shows the average overall cost-growth differential for the RDT&E phase, the procurement phase, and the

---

[10] The changes in cost-growth factor in each joint program were 40 percent to 10 percent in JPATS, 68 percent to 62 percent in F-35, 56 percent to 55 percent in V-22, and 97 percent to 118 percent in JSTARS.

total acquisition phase. This is the average excess cost growth in the joint programs over that in the single-service programs. Acquisition program totals include development, procurement, and in a few cases military construction (MILCON) funding. For each program, the percentage cost growth for the acquisition phase is between that of RDT&E and procurement, as one would expect.[11]

Figure B.1 shows program acquisition cost-growth percentages for the eight programs, side-by-side. Note that the lowest-growth joint program (Joint Primary Aircraft Training System [JPATS] T-6A/B) has only slightly less cost growth than the highest-growth single-service program (T-45 TS). In other words, three of the joint programs experienced higher cost growth than any single-service programs.[12]

In short, the data show that although the total data set is small, there are no outliers and the programs within each set are generally consistent with each other. Thus, the data show that, on average, joint aircraft MDAPs have experienced considerably greater acquisition program cost growth nine years beyond MS B than single-service aircraft programs. This indicates that analysts at MS B have overemphasized the theoretical benefits of joint aircraft programs at MS B and underemphasized the countervailing negative issues, such as increased program complexity, difficulty of meeting multiple-service requirements while maintaining maximum design commonality, and the likelihood of extra weight and unneeded capabilities resulting from the need to accommodate multiple-service requirements and designing to the highest common denominator.

However, in theory, the savings from joint programs do not derive solely from the development and production stages. Significant savings should also accrue from the O&S stage. Appendix C carries out an analysis parallel to that undertaken in Appendix A to determine the maximum amount of O&S savings that could be achieved in an ideal joint fighter program. Appendix D then calculates how much money on average has to be saved in the O&S phase of a joint fighter program to offset the greater average cost growth that historically has taken place on joint aircraft procurement programs compared with single-service programs to result in overall joint LCC savings.

---

[11] For a discussion of different average joint acquisition cost-growth differentials between joint and single-service programs for five years past MS B through eight years past MS B, see Appendix D, pp. 27 and following.

[12] Interestingly, the T-45 TS program, the single-service aircraft with the highest cost growth, had many similarities to a joint program. This is because it was a major modification of a foreign-designed and -developed land-based trainer jet, the British Aerospace Hawk Mk60, designed to Royal Air Force specifications, to make it compatible with Navy carrier operations and other U.S. Navy requirements. Thus, in some respects, it confronted the same types of complexities encountered on a typical joint Air Force/Navy collaborative aircraft development program.

Table B.3
Joint Versus Single-Service Program Acquisition Cost-Growth Differential at Nine Years Past MS B (millions of FY 2012 $)

Program Type	RDT&E	Procurement	Total Acquisition[a]
Single-service programs			
C-17 40 aircraft			
MS B	$6,400	$15,600	$22,300
9 years past MS B	$9,000	$20,000	$29,000
Difference	$2,600	$4,500	$6,800
Percentage	41	29	30
F/A-18E/F 548 aircraft			
MS B	$7,600	$49,700	$57,300
9 years past MS B	$7,450	$48,700	$56,150
Difference	−$150	−$1,000	−$1,150
Percentage	−2	−2	−2
F/A-22 315 aircraft			
MS B	$25,500	$37,300	$63,100
9 years past MS B	$33,000	$43,500	$76,700
Difference	$7,500	$6,200	$13,600
Percentage	29	17	21
T-45 TS 218 aircraft			
MS B	$920	$4,000	$4,900
9 years past MS B	$1,200	$6,000	$7,300
Difference	$280	$2,000	$2,300
Percentage	30	50	47
Single-service programs average cost growth (%)	**25**	**23**	**24**
Joint programs			
JPATS T-6A/B 711 aircraft			
MS B	$423	$3,360	$3,870
9 years past MS B	$368	$5,000	$5,410
Difference	−$55	$1,640	$1,540
Percentage	−13	49	40
F-35 (JSF) 2,443 aircraft			
MS B	$39,300	$155,700	$197,000
9 years past MS B	$59,000	$270,900	$331,000
Difference	$19,700	$115,200	$134,000
Percentage	50	74	68
V-22 Osprey 523 aircraft			
MS B	$4,300	$26,100	$30,600
9 years past MS B	$9,800	$38,600	$47,800
Difference	$5,500	$12,500	$17,200
Percentage	128	48	56

Program Type	RDT&E	Procurement	Total Acquisition[a]
E-8A (JSTARS) 10 aircraft			
MS B	$2,300	$2,100	$4,500
9 years past MS B	$4,800	$3,900	$8,800
Difference	$2,500	$1,800	$4,300
Percentage	107	86	97
Joint programs average cost growth (%)	**67**	**64**	**65**
Joint versus single-service cost-growth differential (%)	**42**	**40**	**41**

[a] RDT&E plus procurement do not sum to total acquisition because total acquisition includes MILCON funding; adjusting to BY 2012 dollars causes inconsistencies because cost-growth percentages for each program were calculated using that program's unique BY dollar calculations, not the inflation indices used in the above table; numbers were rounded to significant digits.

NOTE: Program estimates are adjusted to BY 2012 dollars from each program's MS B BY dollars using Secretary of the Air Force/Financial Management Comptroller Economic and Business Management USAF Raw Inflation Indices (March 5, 2012) based on Office of the Secretary of Defense (OSD) Raw Inflation Rates (February 10, 2012).

Figure B.1

Joint Versus Single-Service Program Acquisition Cost Growth at Nine Years Past MS B

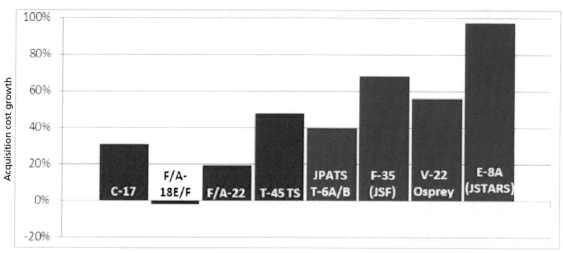

NOTE: Red indicates a single-service program. Blue indicates a joint program.

Appendix C: Calculation of Maximum Joint O&S Savings from an "Ideal" Joint Fighter Program

In Appendix A, we calculated the acquisition cost difference between a notional two-service joint program and two notional single-service programs. In the joint program, two services collaboratively develop a single 100 percent common aircraft and procure it in equal quantities. This was compared with two identical single-service programs, that is, programs in which each service develops and procures the same number of an identical aircraft, with the sum of the two services' aircraft production equal to the total production of the joint program. In this appendix, we make a similar calculation for O&S costs. Again, we assume that the aircraft in each of the two cases are identical and use 100 percent identical support infrastructure, so that the maximum cost savings would accrue. In real programs, there would be less commonality, of course, and less than 100 percent use of common support infrastructure, so the savings would be less.

In this case, because appropriate historical data were available, we took a historical statistical approach. The Air Force Total Ownership Cost (AFTOC) database includes data on O&S costs for a wide range of aircraft for the years 1996 through 2010. The Air Force's Logistics Installations and Mission Support–Enterprise View (LIMS-EV) database includes data on fleet size and flying hours for the same years. We combined these data to estimate how costs change as a function of fleet size, holding flying hours per year constant.

Because we are considering the savings associated with different services acquiring the same aircraft, it is appropriate to include only that part of O&S costs that can conceivably be affected by the existence of the same aircraft in another service's fleet. There are six broad categories of costs included in AFTOC:

1. unit personnel, including flight crews, base maintenance personnel, and other squadron personnel
2. unit operations, which is primarily fuel, plus support services and temporary duty costs
3. maintenance, which includes aircraft consumables, depot-level reparables, depot maintenance and contractor logistics support
4. sustaining support, which includes support equipment replacement and sustaining engineering and program management
5. continuing system improvements, which is modifications
6. indirect support, including installation and personnel support.

We judged that costs incurred in categories 1, 2, and 6 would not be affected by whether other services operated the same aircraft. There is no reason why the number of personnel the Air Force has on a base for each squadron would be affected by whether the Navy operates the same aircraft as well. Thus, personnel, personnel support (including temporary duty), and installation support should not be affected. Nor would Air Force fuel expenses be affected. On the other hand, the other categories clearly have the potential to be affected by the total number of aircraft in operation across all services. The more total aircraft in operation, the larger the number of spare parts (consumables and DLRs), modification kits, and pieces of support equipment produced and managed and the larger the number of overhauls done. Production and management of parts and equipment, and overhaul costs, may well be affected by economies of scale. In addition, a larger force means that the sustaining engineering and contractor management work applies to more aircraft, thus lowering average unit cost. Therefore, in these categories we judged that there was potential for savings as the overall fleet size rose, and we included those costs in our analysis. We refer to them as "scale-related O&S costs" in the rest of this appendix. They account for 63 percent of current expected JSF O&S costs.[1]

To assess how this part of O&S costs would vary with total fleet size, we constructed a data set for the A-10, F-15, F-16, and F-22 from 1996 through 2010.[2] We then did a regression analysis to determine how scale-related O&S costs were affected by total fleet size. Because all the aircraft are Air Force fighters, which are operated and supported by an identical Air Force O&S infrastructure, this approach simulates the O&S cost savings that could be achieved in a joint program made up of common aircraft types across the services using an identical O&S infrastructure.[3]

---

[1] Data are from the JSF Joint Program Office (JPO), from 2011.

[2] Each Air Force fighter type analyzed includes numerous different variants and blocks, some of which differ significantly. For example, not all F-16s in the Air Force inventory are 100 percent identical because of different blocks and variants, such as the F-16A Block 20 or F-16C Block 40. However, our analysis kept the same relative mix of numbers of existing variants of each fighter type when we tested for the effect of changing the size of the force structure on O&S costs for that type of fighter. Thus, we compared the effect on O&S costs of changing the number of common fighter types while keeping constant the existing percentage mix of various variants.

[3] We note that this is a conservative assumption with respect to the benefits of a joint platform, because it assumes that each service benefits equally from an increase in its own aircraft and from an increase in other services' aircraft. In reality, it may be that part of the DLR system, for example, is indeed service-specific, in which there would be lower economies resulting from other services' fleets than from the service's own. This analysis assumes that the joint aircraft is maintained by a truly common supply system in the way that Air Force aircraft in different commands are maintained by a common system. Thus, we are estimating a maximum joint-fleet benefit, much as we did in Appendix A, which considered a maximum joint-fleet acquisition benefit.

Observations were at the aircraft-year-command level, with 1,127 observations in total.[4] That is, each specific aircraft-year-command combination (such as A-10 in Air Combat Command [ACC] in 1996) is a distinct observation. Explanatory variables included

- natural logarithm of number of aircraft (total aircraft inventory [TAI])
- natural logarithm of flying hours per TAI (FH/TAI)
- natural logarithm of empty weight (EW)
- age
- dummy variable for Air Force Materiel Command (AFMC)
- dummy variable for Air Force Reserve Command (AFRC)
- dummy variable for Air National Guard (ANG)
- dummy variable for Air Education and Training Command (AETC)
- dummy variable for Pacific Air Forces (PACAF)
- dummy variable for U.S. Air Forces in Europe (USAFE).

ACC was the excluded command for the dummy variables. The dependent variable was the natural logarithm of scale-related O&S costs. Results are shown in Table C.1. R-square was 0.80, and the standard error of estimate was 0.69. Standard errors were clustered by command to take into account correlation of errors over time within command aircraft.

The results indicate that the coefficient on fleet size is 0.932, which is indeed evidence of economies of scale. This implies that if fleet size doubles, total scale-related O&S costs will increase 91 percent and average scale-related O&S costs will fall by 5 percent.[5] Starting from a point at which the scale-related O&S costs are 63 percent of total O&S costs, a doubling of fleet size will decrease total O&S costs by 2.9 percent.

---

[4] We chose this functional form because there is variation in TAI and FH across commands that can let us more precisely estimate the separate effects on cost of these two variables. Given this functional form, if we increase the number of aircraft at each command by a constant percentage (which means, of course, increasing the entire fleet by that percentage), then total costs will go up as indicated by the coefficient on TAI in the equation. That is precisely what we are looking for: How does cost respond to fleet size when nothing else changes (such as FH/TAI and command distribution of the aircraft). To increase our confidence in the results, we also ran the model without the command disaggregation. The coefficient on TAI was almost identical. This is discussed further below when we discuss alternative functional forms.

[5] 2 raised to the 0.932 power is 1.91, hence the 91 percent increase in total scale-related cost. 1.91/2 is 0.95, hence the 55 percent decrease in average scale-related cost.

Table C.1
Results of Scale-Related O&S Cost Regression

Explanatory Variable	Estimated Coefficient	Standard Error	t-statistic
ln TAI	0.932	0.03	32.3
ln FH/TAI	0.63	0.12	5.1
ln EW	0.89	0.42	2.2
Age	0.018	0.009	2.2
AFMC dummy	0.45	0.28	1.6
AFRC dummy	−0.58	0.16	−3.6
ANG dummy	−0.40	0.10	−4.1
AETC dummy	−0.25	0.11	−2.3
PACAF dummy	−0.26	0.11	−2.3
USAFE dummy	−0.31	0.13	−2.5

We also considered other functional forms for this relationship. As discussed in footnote 4, we left out the command dummies to ensure that we were indeed capturing fleet-wide economies of scale. The coefficient on ln TAI was 0.94, with a standard error of 0.014. Thus, its confidence interval overlaps that of the Table C.1 regression. We also added Mission Designation Series fixed effects and FY fixed effects, rather than use EW and age as explanatory variables. These resulted in a coefficient on ln TAI of 0.81, with a standard error of 0.06. Its confidence interval also overlaps that of the Table C.1 regression. We also used a functional form that included the MDS fleet in other commands as well as the MDS fleet in the given command. The coefficient on the MDS fleet in other commands was very small (0.014) with a confidence interval including zero (s.e. = 0.017). This again gives us confidence that the Table C.1 regression captures fleet-wide economies of scale. Finally, we added a [ln TAI]-squared term to represent a case of a nonconstant elasticity. This resulted in an elasticity estimate of 1.03 (at average TAI level), with a standard error of 0.036. This specification implies slightly decreasing economies of scale; the confidence interval includes increasing economies of scale.

Different functional forms give different point estimates of the elasticity with respect to fleet size, but all confidence intervals include the possibility of increasing returns to scale. The specification with the highest fleet-wide economies of scale is the MDS fixed-effects and fiscal year fixed-effects versions. It implies that if fleet size doubles, total scale-related O&S costs will increase 75 percent and average scale-related O&S costs will fall by 12 percent. Starting from a point at which the scale-related O&S costs are 63 percent of total O&S costs, a doubling of fleet size will decrease total O&S costs by 7.8 percent. This, as with the 2.9 percent estimate from the Table C.1 regression, is less than our base estimate of the O&S savings required to offset the joint acquisition cost-growth premium (see Table D.2). We originally chose the Table C.1 specification as our preferred equation because of previous cost-estimating work. These sensitivity results show that one finds differences in point estimate values, not in kind, using different functional forms. Our preferred specification results in a scale economy estimate that is in the middle of the range found using other functional forms.

Appendix D: Exploring the Magnitude of Joint O&S Savings Needed to Offset Joint Acquisition Cost-Growth Premium

The joint aircraft acquisition cost-growth premium described in Appendix B might be offset by O&S savings enjoyed by the joint program over those of two or more single-service programs. Quantifying the O&S savings needed for this offset required the development of an arithmetic model. Using the model, with inputs including the cost-growth premium in joint programs at nine years past MS B, resulted in the calculated 10.3 percent of O&S costs that a joint program must save (as shown in the main report) to offset the 41 percent joint program acquisition cost-growth premium (as shown in Table B.3).[1]

The output of the model is the percentage of joint O&S cost savings required to overcome a given level of higher cost growth in joint acquisition programs over single-service acquisition programs. Below, we discuss the inputs that result in the O&S savings that we estimate are required to offset the average joint program acquisition cost-growth premium and the origins of those inputs. We then assess the range of plausible values for each input.

The 10.3 percent required O&S cost savings is based on conservative input values. Several key inputs with significant uncertainty were specified to give the joint program the maximum potential benefit. This was done to ensure that we do not underestimate the potential advantages of joint program execution.

The model consists of the following key variables and assumed values (explained below in this appendix):

- SS RDT&E CG: This is the average single-service program cost growth in RDT&E at nine years past MS B measured in the four programs analyzed in Appendix B: 25 percent.
- SS procurement CG: This is the single-service program cost growth in procurement at nine years past MS B measured in the four programs analyzed in Appendix B: 24 percent.
- Joint RDT&E CG: This is the average joint program cost growth in RDT&E at nine years past MS B from the four programs analyzed in Appendix B: 68 percent.

---

[1] The nine-year point past MS B was selected because the JSF program was approximately nine years past its MS B as of its December 2010 SAR, which was the latest SAR available at the time this work was completed in December 2011.

- Joint procurement CG: This is the average joint program cost growth in procurement at nine years past MS B from the four programs analyzed in Appendix B: 64 percent.
- SS RDT&E fraction: This is the fraction of the total acquisition that is RDT&E funding in each single-service program as estimated at MS B. It is based on historical fighters programs with a substantial production run. We use a representative value of 20 percent.
- SS procurement fraction: This is the fraction of the total acquisition that is procurement funding in each single-service program as estimated at MS B. It is based on historical fighters programs with a substantial production run. We use a representative value of 80 percent.
- SS O&S CG: This is the single-service program cost growth in O&S as measured from MS B. For this calculation we assume 0 percent. The specified value is neither conservative nor representative. It provides the maximum variance in potential outcomes.
- Joint RDT&E savings: This is the RDT&E savings in a joint program compared with the cost of two single-service programs as estimated at MS B. We use the maximum theoretical value of 50 percent as discussed in Appendix A.
- Joint procurement savings: This is the procurement savings in a joint program compared with the cost of two single-service programs. The value of 13 percent (described in Appendix A) is both representative and conservative at the same time. It assumes an average cost improvement curve (CIC) of 87 percent, which is representative. It also assumes 100 percent commonality between all variants and 0 percent commonality in all single-service variants. This combination gives the maximum theoretical advantage to joint programs.
- Joint MS B O&S fraction: This is the portion of the joint program's LCC estimated in O&S at the program's MS B. The 46 percent value is considered representative.[2]
- Joint MS B O&S savings: This is the O&S savings for the joint program over that for the two single-service programs estimated at MS B. The theoretical maximum value of 2.9 percent is used (see Appendix C).

Model Variable Values and Sensitivity Analyses

Joint Acquisition Cost-Growth Premium

The 41 percent joint program acquisition cost-growth premium (discussed in Appendix B) is expressed in four variables in the model: SS RDT&E CG, SS procurement CG, joint RDT&E CG, and joint procurement CG. The mathematical relationships showing how these four variables equal the joint program acquisition cost-growth premium can be inferred from Table B.3.

---

[2] This is the percentage of O&S costs compared with total program costs for JSF at MS B.

Note that the model implicitly assumes zero acquisition dollars in both MILCON and acquisition-related O&M for both joint and single-service programs.

Acquisition Cost Split

The single-service program split of acquisition dollars between RDT&E and procurement is determined by the complexity of the aircraft and the number of production aircraft. For a tactical fighter program with a substantial production quantity, development can be as low as 12 percent and as high as 33 percent of acquisition costs. The low end of the range is for a derived fighter design such as the F/A-18E/F; the high end is for a highly complex fighter design. The percentage in RDT&E can be even higher if production is cut short and only a few hundred fighter aircraft are built. When this occurs, coupled with a highly complex design, RDT&E can be as much as 60 percent of total acquisition. This occurred in the F-22 program.

Table D.1 shows our base case variable values in the left-most data column. The effects of different single-service RDT&E/production value splits are shown in the other columns. When RDT&E exceeds 33 percent (holding all other variables constant), it appears that joint programs make sense, because the required savings in joint O&S costs to overcome a 41 percent acquisition cost growth premium turn negative.

Table D.1
Sensitivity to Acquisition Distribution Between RDT&E and Procurement (in percent)

	Base Case	Low RDT&E	High RDT&E	High RDT&E; Low Quantity
SS RDT&E	20	12	33	60
SS procurement	80	88	67	40
Required savings in joint O&S to overcome 41% joint acquisition cost-growth premium	10.3	16.6	−1.0	−29.8

Growth in O&S Costs

Up until ten years ago, O&S costs were not consistently estimated and reported in SARs. Useful O&S data at the time of MS B were found in only a few of the programs in our eight-program data set. For this reason, we assumed 0 percent cost growth for the SS O&S CG variable. Because O&S costs reflect the complexity of a fighter aircraft, and the underestimate of acquisition costs often reflects an underestimation of a system's complexity, one might expect O&S costs to increase in percentages similar to those of acquisition estimates.[3] Given this, 50 percent or more in O&S cost growth from MS B in a single-service program is a reasonable assumption. Because of the high uncertainty for the appropriate value of the SS O&S CG variable, and because increasing the value of this variable reduces the range of the model's outcomes (when other key inputs are varied), we chose to specify this variable at 0 percent to explore the range in potential outcomes. Table D.2 shows that as O&S costs grow (first data row), the percentage savings required to offset the 41 percent joint acquisition cost-growth premium moves towards zero (last row). This is because O&S costs become a larger fraction of total LCC in both single- service (second data row) and joint programs (third data row) as the SS O&S CG value increases.

Table D.2
Sensitivity to SS O&S CG Variable Value (in percent)

	Base Case	Reasonable	High	Extreme
SS O&S cost growth	0	50	100	200
O&S fraction (after cost growth) of LCC in SS programs	36	46	53	63
O&S fraction (after cost growth) in joint programs	32	43	50	61
Required savings in joint O&S to overcome 41% joint acquisition cost-growth premium	10.3	6.9	5.2	3.4

Specifying Joint RDT&E and Procurement Savings

The 50 percent joint RDT&E savings is an absolute maximum because it assumes that the aircraft in each of the two single-service programs are identical in design, size, and complexity to the single aircraft design in the joint program, yet at the same time the two single-service aircraft have zero production commonality. This is highly unlikely with two simultaneous tactical fighter acquisition programs. The single-service aircraft would almost certainly share some avionics, subsystem components, exotic materials, and possibly engines, as well as production of these items, so that economics of scale could be achieved. A more realistic theoretical value of saving in RDT&E is 30 percent to 40 percent. Assuming this savings at 50 percent is highly conservative in favor of joint programs. Because the model is highly sensitive to this input, assuming a lower percentage for this value substantially increases the O&S savings required to offset the joint aircraft cost-growth premium. This is illustrated in Table D.3.

---

[3] Acquisition programs typically experience from 0 percent to 200 percent cost growth, with a median value of about 50 percent.

Table D.3
Sensitivity to Joint RDT&E Savings Variable Value (in percent)

	Base Case/Maximum	Optimistic	Likely
Joint RDT&E savings	50	40	30
Required savings in joint O&S to overcome 41% joint acquisition cost-growth premium	10.3	14.7	19.0

The 13 percent joint procurement savings is also an assumption very favorable to joint programs. It assumes equal numbers of aircraft acquired in both single-service programs. If a difference in quantities exists, then savings are reduced. The larger the difference in the quantities of the two or more participants, the lower the actual savings. This effect is small, but it is real if any cost improvement with quantity occurs.

The 13 percent joint procurement savings also assumes zero commonality between single-service aircraft production, and 100 percent commonality of design, components, and production between joint aircraft . In reality, more than 80 percent commonality in joint variants' design and components is difficult to achieve. Seventy to 90 percent was the goal in the JSF program, but it is not likely to be achieved. In addition, at least 10 percent to 20 percent or more commonality in single-service variants (with each other and with other nonvariant aircraft) is plausible, if for example, two different single-service fighters use the same or similar engine, as was the case with the F-111 and the F-14, and the F-15 and F-16.[4] As a result of these two realities, the actual additional commonality leverage achievable in a joint program versus a single-service program is closer to 50 percent or less, not the 100 percent we assumed in the base case. Table D.4 shows the effect on required savings in joint O&S to overcome the 41 percent joint acquisition cost-growth premium.

---

[4] The rule of thumb factor is that engines typically account for about one-third of the production costs of fighter aircraft.

Table D.4
Sensitivity to Commonality Differential

	Base Case/Maximum	Optimistic	High End of Likely	Low End of Likely
Commonality differential between joint and single service (%)	100	80	60	50
Cost-quantity improvement curve (%)	87	87	87	87
Production quantity split	50/50	50/50	50/50	50/50
Required savings in joint O&S to overcome 41% joint acquisition cost-growth premium (%)	10.3	14.8	19.1	21.1

An estimated 87 percent cost-quantity improvement curve for acquisition throughout procurement is typical for fighter aircraft at the time of their MS B. Cost-quantity improvement curves mostly fall between 85 percent and 90 percent in fighter aircraft programs with substantial production quantities. As shown in Table D.5, steeper curves require lower joint O&S savings than do flatter curves to overcome the 41 percent average joint acquisition cost-growth premium.

Table D.5
Sensitivity to Cost-Quantity Improvement Curve

	Base Case	Optimistic	Pessimistic
Cost-quantity improvement curve (%)	87	85	90
Commonality differential between joint and single service (%)	100	100	100
Production quantity split	50/50	50/50	50/50
Required savings in joint O&S to overcome 41% joint acquisition cost-growth premium (%)	10.3	6.7	15.5

O&S Fraction of LCC

The joint MS B O&S fraction value of 46 percent is taken from the F-35 program at its MS B. As a rule of thumb, O&S is reasonably estimated at about half of LCC in aircraft programs at MS B. This percentage varies with a number of program characteristics and programmatic assumptions, the most important of which are weapon system type and the number of years of operation assumed for each unit. F-35 estimates assume 30 years of operations per aircraft, which is appropriate for tactical aircraft built in the past few decades. Reasonable estimates of minimum and maximum values are

33 percent and 67 percent, respectively. Table D.6 illustrates that higher values for the joint MS B O&S fraction variable require lower savings in joint O&S to overcome the 41 percent joint acquisition cost-growth premium, and vice versa.

Table D.6
Sensitivity to Joint MS B O&S Percentage of LCC

	Base Case	Minimum	Maximum
Joint MS B O&S	46	33	67
Required savings in joint O&S to overcome 41% joint acquisition cost-growth premium	10.3	25.3	6.1

Joint Program MS B O&S Savings

The 2.9 percent value joint MS B O&S savings is considered a reasonable maximum for a two-variant joint aircraft program. (See Appendix C for a detailed explanation for determining the value of this variable for two-variant joint programs.) The reasons that this value is a reasonable maximum are similar to those outlined for the joint RDT&E savings and joint procurement savings variables. We can find no evidence that the benefits of joint O&S are explicitly quantified in the cost estimates made at MS B. However, qualitative assertions of such benefits are typically made at the time of these estimates. It is likely that some quantitative benefit is included in MS B O&S estimates through assumptions regarding economies of scale associated with the larger total aircraft quantity inherent in a joint program. With all of the above in mind, a more plausible value for savings in O&S for a two-variant joint program is 1 percent to 2 percent.

Accounting for Uncertainty in the Joint Program Acquisition Cost-Growth Premium

The 41 percent acquisition cost-growth premium in joint aircraft programs at nine years past MS B is calculated from averages at that point in time. However, the magnitude of this premium varies from year to year as shown in Figure 2.1 in the main report and in Table D.7. It is necessary to account for this variation when calculating the amount of O&S savings required to offset the acquisition cost-growth premium.

Costs tend to grow in acquisition programs as they execute; thus, it is reasonable to infer that the acquisition cost-growth premium measured in percentage points in joint programs is partially a function of the point in time past MS B at which it is measured. The data in the top row of Table D.1 support this inference. To remove this temporal effect, we express the acquisition cost-growth premium as a ratio of cost-growth factors at each point after MS B rather than as percentage points. Put more specifically, because this premium can be calculated at multiple points after a milestone with different average values for both the joint and single-service program data sets, and because the value of the premium differs from year to year, the raw percentage differences in the averages at any one point in time after the milestone is not sufficient to calculate the O&S savings range required in joint programs to offset the acquisition cost-growth premium. The calculations in Table D.7 transform the joint cost-growth acquisition premium from a percentage point value specific to a point in time after MS B to a ratio that is applicable to any point in time after MS B. Calculating these ratios at all relevant points in time provides a range of estimate for the joint program acquisition cost-growth premium.

The calculation works as follows: In the year 9 column on the far right of the table, the single-service cost-growth average is 1.24 or 24 percent growth (1.00 represents the MS B estimate and thus no cost growth). At that same point after MS B, the joint cost-growth average is 1.65, or 65 percent growth. Subtracting the 24 percent from the 65 percent gives our base-case joint aircraft cost-growth premium of 41 percent. The ratio of the cost-growth averages, 1.65/1.24, is 1.333. The cost-growth ratio represents the premium in joint programs. The ratios in the bottom row of Table D.7 show a minimum of 1.249 in year 7 and a maximum of 1.364 in year 8. These ratios define the lower and upper bounds of the band of uncertainty in the joint program acquisition cost-growth premium.

Table D.7
Calculation of Joint Program Acquisition Cost-Growth Premium Ratios

Year Past MS B	5	6	7	8	9
Average joint program acquisition cost-growth premium (%)	30.2	30.2	31.1	43.8	41.2
Joint program - average total program acquisition cost-growth factor	1.43	1.48	1.56	1.64	1.65
Single-service program - average acquisition total program cost-growth factor	1.13	1.17	1.25	1.20	1.24
Joint program over single-service acquisition program ratio	1.268	1.257	1.249	1.364	1.333

To express these boundaries in our model, we develop similar ratios for RDT&E and procurement using year 7 data for minimums and year 8 data for maximums. These ratios, along with those for year 9 data, are shown in Table D.8. The minimum and maximum ratios can be applied to the single-service program average cost-growth averages for RDT&E and procurement, respectively, in any year after MS B to calculate the minimum and maximum values for RDT&E and procurement joint program cost growth, thus making it possible to quantify the uncertainty in the O&S joint program cost-growth savings required to offset the acquisition cost-growth premium.

To calculate the O&S savings range required to offset the acquisition cost-growth premium, we fix the cost growth in the single-service programs using their average values at year 9 and then apply the minimum and maximum of the ratios of the joint cost-growth factors over the single-service cost-growth factor from the years for which data are available (years 5 through 9).

Joint Program Acquisition Cost-Growth Premium Minimum and Maximum Values

In Table D.8, single-service program average cost growth at year 9 is used with the ratios from that year, the minimum year, and the maximum year. The joint program cost-growth estimates, calculated using the appropriate ratios, are also shown. All other variables from our model remain as shown at beginning of this appendix.

At the bottom of Table D.8 are the resulting O&S savings required to offset the joint acquisition cost-growth premium using data from year 9 and the results of applying the minimum and maximum ratios to year 9 single-service cost-growth averages. Using year 9 values alone, the joint program must save more than 10.3 percent in O&S costs in comparison with two single-service programs to save money overall.

Table D.8
Required Joint O&S Savings Based on Joint Acquisition Cost-Growth Minimums and Maximums

Variable/Model Outcome	Year 9 Values	Minimum: Year 7 Values	Maximum: Year 8 Values
Single-service program average RDT&E cost-growth percentage—year 9 (%)	25	NA	NA
Single-service program average procurement cost-growth percentage—year 9 (%)	24	NA	NA
Total acquisition joint program cost-growth premium ratio	1.333	1.249	1.364
O&S savings required to offset joint acquisition cost-growth premium (%)	10.3	−5.1	15.4

Using the *minimum* joint cost-growth acquisition premium ratio, the joint program could experience up to an *additional* 5.1 percent in O&S costs and still be less expensive than the two single-service programs. Using the *maximum* joint cost-growth acquisition premium ratio, the joint program must save 15.4 percent in O&S costs in comparison with two single-service programs. The plausible maximum joint O&S cost savings of 2.9 percent (see Appendix C for details) is roughly in the middle of this range. Our finding using the year 9 joint acquisition cost-growth premium of 41 percent showing that more than 10.3 percent savings in O&S costs must be achieved by a two-aircraft joint program compared with two single-service aircraft lies well within the range of this uncertainty band.

Clearly there is much uncertainty as to whether the joint program acquisition cost-growth penalty, the size of which itself is uncertain, can be overcome by joint O&S cost savings. The large range for O&S costs potentially needed to offset the average joint program acquisition cost-growth premium, the small sample size of four each joint and single-service programs, and the high uncertainty of several high-leverage variables contribute to this uncertainty.

However, our specification of the joint RDT&E savings and joint procurement savings variables at their plausible maximum, and the calculation of the plausible maximum theoretical O&S savings possible from an "ideal" joint fighter program as shown in Appendix C, suggest that a two-variant joint aircraft program is unlikely to save money compared with two comparable single-service aircraft programs.

Appendix E: Primary Methodology for Comparing JSF Costs with Those of Three Notional Single-Service Fighters

This appendix describes how we calculated the cost of three separate notional single-service programs that might have been undertaken instead of the JSF program.[1] We first describe the initial projected cost of the JSF program in 2001. We then calculate what the cost of three notional single-service programs would have been in the absence of any cost growth and contrast that with the initial projected cost of the JSF (i.e., also without cost growth). We then calculate what the cost of the single-service programs would have been if they had experienced the same cost growth as the F-22. We use the F-22 experience as our analogue for the cost growth that would be expected of a single-service advanced technology fighter-attack aircraft.[2] We then contrast the cost of our projected single-service programs, after cost growth, with the current expected JSF cost, which includes the cost growth that it has experienced.

Baseline JSF Cost Estimates

We begin by characterizing the projected cost of the JSF as of December 2001, which we take as the baseline for this study. (All dollar figures in this appendix reflect 2002 purchasing power, or FY 2002 dollars.) All of the December 2001, or initial, estimates for the JSF are from its December 2001 SAR. The total R&D cost for all three variants was $32.4 billion.[3] We include only costs to the United States in our analysis, so we exclude the $2.2 billion R&D contribution from the UK. (This is identified as "RDT&E – Other" in the SAR. It has remained unchanged since.) Thus, the expected R&D cost of all three variants to the United States was $30.2 billion.

---

[1] Both this appendix and Appendix F support the analysis in the main report in Chapter Three, especially sections 1 and 2.

[2] We view this as a conservative approach because of the different technical challenges faced by the two programs. The F-22 is a much higher-performance fighter than the JSF Air Force variant. It was a relatively high-risk development program because it was the first supersonic/super cruise low observable (LO) fighter ever developed, and the first radar-equipped LO fighter using the world's first active electronically scanned array antenna for fire control. It was also the first fighter with a fully integrated avionics system. However, its cost growth was below the average of the single-service programs from Appendix B. In the Sensitivity Results section below, we show the sensitivity of our results to the projected single-service programs' cost growth.

[3] Note that throughout this appendix, in both the tables and the text, numbers may not sum exactly because of rounding.

31

The expected production program in the December 2001 SAR included 1,763 Air Force aircraft and 1,089 Department of the Navy (DON) aircraft. The DON aircraft were split between the 480 CV (Navy) and 609 STOVL (Marine Corps) variants. The total production cost of the Air Force variant was $80.5 billion and of the DON variants $64.6 billion. The split of the DON cost between the CV and STOVL variant was not given. According to other sources, the expected recurring unit flyaway cost of the CV variant, in FY 2002 dollars, was $48.1 million and of the STOVL variant, $46.0 million.[4] We estimate that the total production cost of the 480 CV variants was $29.2 billion and of the 609 STOVL variants, $35.4 billion. (That is, we split the total cost between the two variants so that the ratio of total cost per aircraft for the two was equal to the ratio of the recurring unit flyaway cost. This is equivalent to assuming that the ratio of total production cost to flyaway cost is the same for the two.) As stated in the December 2001 SAR (p. 41), JSF plans at the time called for 150 UK STOVL variants to be produced, as well as the 609 Marine Corps aircraft. The $35.4 billion production cost includes only the cost to the United States.

The December 2001 SAR stated that the O&S costs to the United States of the JSF would be $152 billion. There was no information about comparative CTOL (Air Force), CV, or STOVL cost, although the SAR does explicitly state, "'Total O&S Cost' below reflects the O&S costs for all three variants."

Baseline Single-Service Aircraft Cost Estimates

We now turn to our estimate of the cost of three notional single-service programs that might have been undertaken instead of the JSF.

For R&D, we accepted the JSF JPO and contractor's estimate that JSF would be able to do the RDT&E of three different fighter variants for 60 percent of the cost of three single-service RDT&E programs. Thus, we assumed that RDT&E for three single-service programs would be 67 percent higher than the December 2001 SAR RDT&E estimate.[5] This leads to a total R&D cost for the three single-service programs of $54.0 billion. We

---

[4] Data are from the F-35 JPO .

[5] JPO and contractor estimates varied over time. In 1998, three years before to MS B, JPO documentation said that RDT&E for three single-service programs would be twice the JSF RDT&E costs. By the early 2000s, this had been adjusted to the estimate that RDT&E for three single-service programs would be 1.5 times the JSF RDT&E costs. One publicly available JPO briefing from 2008 shows that RDT&E for three single-service programs would be 1.67 times JSF RDT&E costs, which is the figure we use in our analysis. See, for example, the briefing entitled *JSF Production*, presented by then Major General Charles R. Davis, then Program Executive Officer, F-35 Program Office, at the Aviation Week Aerospace and Defense Finance Conference, New York, November 2008. We note that the "theoretical maximum" value for this ratio, according to Appendix A, would be 3.0 but that would be relevant only for identical variants, which were, of course, never considered.

assume that the UK contribution (for the STOVL program in this case) would have remained constant at $2.2 billion, so that the U.S. cost would have been $51.8 billion. (This is in contrast to our approach in Appendix A, which showed the maximum possible benefit of commonality. In the maximum possible case, the JSF would be able to do the RDT&E of three different fighter variants for 33 percent of the cost of three single-service RDT&E programs—each of which would be identical, of course.)

To estimate the production cost of three single-service programs, we must remove the benefits of shared learning from the 2001 JSF forecast. We do this by estimating a cost-improvement relationship for each variant. (For reference, we show the December 2001 SAR RDT&E production level and unit cost estimate, by year, at the end of this appendix.) In a cost-improvement relationship, the cost of each production aircraft is a function of the cumulative production level to that point. That is, the cost of the nth aircraft produced is a function of n. Because of economies of learning, cost falls with n. The December 2001 JSF SAR shows the numbers expected to be produced in each year and the production cost incurred in each year. (Only the combined DON quantity and cost are shown in the SAR. We estimate individual CV and STOVL production quantities by applying the 480/609 CV/STOVL production split to each year's total DON quantity. We estimate the annual cost of each year's CV and STOVL production lot by applying the 29.2/35.4 total cost split to each year's total DON cost.)

To estimate the cost-quantity improvement curve for each variant, we must determine the appropriate concept for "cumulative production" for each. In a single-service program, this is straightforward: It is simply the cumulative number of units of the aircraft produced. If the JSF variants were all identical, it would also be straightforward for the JSF joint program: It would be the cumulative number of all aircraft produced. The JSF variants are not identical, however. Therefore, for each variant, the appropriate cumulative production number should include only that part of the other variants that is common with the given variant. As discussed in the main report, the goal of DoD was to achieve 80 percent commonality among the variants. In this analysis we use that figure and thus represent each U.S. variant as being 80 percent common with the other two. Analogously, we also represent the U.S. and foreign versions of each variant (e.g., the U.S. and UK STOVL variants) as being 80 percent common with each other and the U.S. version of one variant and the foreign version of another (e.g., the U.S. CTOL and the UK STOVL variants) as being 64 percent common. For example, we represent the U.S. CTOL and the U.S. STOVL variants as being 80 percent common, the U.S. STOVL and the foreign STOVL as being 80 percent common, and the U.S. CTOL and the foreign STOVL as being 64 percent common. (This is also in contrast to our approach of Appendix A, which showed the maximum possible benefit of commonality. In the maximum possible case, there would be 100 percent commonality among all variants.)

33

Given this commonality relation, we can calculate an "effective cumulative production" variable for each U.S. variant. For example, for the U.S. CTOL, the effective cumulative production would be U.S. CTOL cumulative production plus 80 percent of U.S. cumulative CV and STOVL production plus 80 percent of foreign cumulative CTOL production plus 64 percent of foreign cumulative CV and STOVL production. Using these effective cumulative production variables, we estimate a learning curve for each variant of the form

$$AC(t) = A\left[ECP(t)\right]^{\gamma}.$$

In this equation, $AC(t)$ is the unit cost of the variant in time period t, and $ECP(t)$ is the effective cumulative production level in that period. A is a parameter that can be interpreted as first-unit cost, and γ is a parameter that is related to the learning rate (λ) through the relation

$$\lambda = (2)^{(\gamma)}.$$

As a practical matter, for "effective cumulative production" in each time period, we use the average of (1) effective cumulative production through the time period in question, and (2) effective cumulative production through the previous time period. This is the same as using effective cumulative production through the midpoint of the year in question. Also as a practical matter, we estimate the learning curve shown above in logarithmic form, so that we get estimates of $\ln(A)$ and γ. Results are shown in Table E.1.

Table E.1
Regression Results for JSF Variant Learning Curves

	CTOL	CV	STOVL
ln A	5.48	5.64	5.61
s.e. (ln A)	0.03	0.03	0.03
A (first-unit cost in millions of 2002 $)	239	283	272
γ	−0.24	−0.24	−0.24
s.e. (γ)	0.004	0.004	0.004
Learning rate (%)	84.8	84.8	84.8
s.e.e	0.04	0.03	0.03
R-square	0.99	0.99	0.99

We then calculate what the cost of each variant would be if it were produced in a stand-alone single-service program rather than in the JSF program. We do this by recalculating the cost of each production lot using only the cumulative production level of the specific variant and not including cumulative production of other variants. We note

34

that we still include the effect of shared learning with the foreign version of each variant. Thus, the Marine Corps STOVL version does get the benefit of 80 percent shared learning with the UK version. We also note that the initial unit cost is assumed to be the same in both the single-service and joint programs.

For example, the level of cumulative production for the U.S. CTOL variant only is 1,763. In contrast, the level of effective cumulative production for the U.S. CTOL, which includes all the shared learning from other U.S. variants and the UK STOVL, is 2,730. Thus, in our estimate of what a U.S. stand-alone single-service CTOL program would cost, we give only the learning benefit from 1,763 rather than from 2,730. Table E.2 shows, for each variant, average unit cost and effective cumulative production for both the single-service and JSF programs.

Table E.2
Effective Cumulative Production and Average Unit Cost for Single-Service and JSF Programs

	CTOL		CV		STOVL	
	Single-Service Program	JSF Program	Single-Service Program	JSF Program	Single-Service Program	JSF Program
Effective cumulative production	1,763	2,730	480	2,474	609	2,523
Average production cost (millions of 2002 $)	52.6	45.7	85.0	60.8	77.3	58.2

Table E.2 shows that the CV and STOVL versions suffer the most in the stand-alone programs, because the loss of joint learning is much greater on a percentage basis for the two. The CV variant costs 28 percent less in the joint than in the single-service program and the STOVL variant 25 percent less. The CTOL version, in contrast, costs 13 percent less. Put another way, the CV and STOVL versions gained substantially from the joint program, because of the relatively small sizes of their production totals, and the CTOL not so much, because it was still the bulk of the production program.

We now turn to the single-service O&S costs. Appendix C describes the historical statistical relation we derived between scale-related O&S costs and total fleet size: Scale-related O&S costs are proportional to fleet size raised to the power 0.932, which implies that scale-related O&S costs per aircraft are proportional to fleet size raised to the power −0.068. Appendix C also points out that only part of total O&S costs is related to fleet size—63 percent in the JSF case. According to the 2001 SAR, the O&S costs per aircraft were $53.3 million. To assess the effect on O&S costs of three single-service programs versus the JSF program, we employ the following relation:

$$O\&S = B[TAI]^{\delta} + C.$$

In this relation, *O&S* is total O&S costs per aircraft, and *C* is that part of O&S costs that is not dependent on fleet size. Scale-related O&S costs are represented by the expression $B[TAI]^{\delta}$, in which *TAI* is total fleet size, and *B* and δ are parameters. Using the statistical analysis of Appendix C, the value of δ is –0.068. Because 37 percent of O&S costs is not related to fleet size, we take *C* to be 19.7 and *B* to be 57.6. Given the 2001 JSF total fleet-size estimate of 2,852, the above relation leads to per aircraft O&S costs of $53.3 million, as given in the SAR.

To estimate what O&S costs would be in single-service programs, we recalculate the O&S cost relation for each variant using the fleet size of that variant rather than the joint fleet size of 2,852. Table E.3 shows the resulting O&S costs per aircraft.

Table E.3
O&S Costs per Aircraft for Single-Service and JSF Programs

	Joint	CTOL	CV	STOVL
Fleet size (aircraft)	2,852	1,763	480	609
O&S costs per aircraft (millions of 2002 $)	53.3	54.4	57.6	57.0

Total O&S costs are $152 billion in the joint case and $158 billion in the single-service case, so the joint program saves 4 percent over single-service programs. Again, the CV and STOVL programs enjoy a larger benefit than the CTOL because of the relative size of the single-service programs.

Table E.4 summarizes the results of our comparison of the JSF program with three single-service programs, all in the absence of cost growth. The figures in the JSF column are directly from the December 2001 SAR.

Table E.4
Comparative Cost of JSF Versus Three Single-Service Programs, Before Cost Growth
(billions of 2002 $)

	JSF	Three Single-Service Programs
R&D	30.2	51.8
Production	145.1	180.5
O&S	151.9	158.2
Total	327.3	390.6

Our estimate is that, in the absence of cost growth, the JSF would have saved 16 percent of total cost versus three single-service programs.

JSF Estimates with Cost Growth

We now turn to a comparison of the costs of the JSF and three single-service programs after cost growth. We first note that the number of aircraft in the program has changed, as shown in Table E.5.[6]

Table E.5
Number of Aircraft in JSF Program in 2001 and 2010

Aircraft	Number in 2001	Number in 2010
U.S. CTOL	1,763	1,763
U.S. CV	480	340
U.S. STOVL	609	340
Foreign CTOL	0	535
Foreign CV	0	0
Foreign STOVL	150	195
Total	3,002	3,173

The change in the number of aircraft in the JSF program since 2001 implies that simply comparing the total cost estimated in 2001 with the current estimated total cost is not appropriate. (By "current estimated cost," we mean the estimate of the December 2010 SAR, which was the last one available for analysis in this study.) Instead, we must adjust the current estimated cost to what it would have been had the number of aircraft in the program stayed the same. Comparing this cost-adjusted-to-2001-number-of-aircraft

---

[6] Data are from the F-35 JPO, from 2010.

estimate with the 2001 cost estimate gives us an appropriate measure of cost growth in the JSF program. Likewise, we apply the cost-growth factors for single-service programs holding number of aircraft constant at the December 2001 level. Thus, we get an appropriate comparison of the JSF and single-service programs after cost growth.

The December 2010 SAR shows that total R&D costs (in 2002 dollars) are now estimated to be $48.4 billion. For this assessment, we assume that the UK contribution would have stayed constant at $2.2 billion, so that the costs to the United States would have been $46.2 billion. Compared with the 2001 costs to the United States of $30.2 billion, this is a 53 percent cost growth.[7]

We estimated what JSF production costs would have been had the number of aircraft in the program not changed with the following method. Using the cost-improvement regressions presented above, we recalculated what JSF costs would have been (before cost growth) at the new production levels. The new cost calculations include both the loss of learning resulting from the decrease in production of U.S. aircraft and the gain in learning resulting from the increase in production of foreign aircraft. Table E.6 shows the steps in the calculations. In the CTOL case, the 2001 SAR gave a cumulative production level of 1,763, and we calculated a cumulative production level of 2,730, as described above. This leads to a unit cost of $45.7 million, and a total cost of $80.5 billion. After removing the lost CV and STOVL aircraft and adding the additional foreign aircraft, the 2010 program (which still has 1,763 CTOL aircraft) has practically unchanged effective cumulative production, at 2,860. This small increase in effective cumulative production slightly lowers unit cost. For the CV and STOVL programs, the results are somewhat different. Although the effective cumulative production is about the same as in the 2001 case, the CV and STOVL programs end in 2014 in the 2010 SAR, whereas the CTOL program extends to 2026. Thus, they miss the benefit of the 2014–2026 CTOL production learning economies, so their unit cost increases somewhat.

---

[7] We note that the actual foreign contribution to RDT&E is now $4.4 billion, but we assess that the additional contribution would not have occurred in the absence of new foreign partners, whose addition has increased the size of the foreign program from 150 to 730. One could argue that part of the increased RDT&E cost is due to the additional partners and so should be removed from our estimate. Given the emphasis on commonality in the JSF program, we judged that this was not appropriate. Had we assumed, instead, that the U.S. cost would have been net of the entire $4.4 foreign contribution, our estimate of RDT&E cost growth would have been 47 percent instead of 53 percent.

Table E.6
Cumulative Production, Effective Cumulative Production, and Cost in the JSF Program in 2001 and 2010

	CTOL	CV	STOVL
2001 cumulative production	1,763	480	609
2001 effective cumulative production	2,730	2,474	2,523
Unit cost of 2001 program before cost growth (millions of 2002 $)	45.7	60.8	58.2
Total cost of 2001 program before cost growth (billions of 2002 $)	80.5	29.2	35.5
2010 cumulative production	1,763	340	340
2010 effective cumulative production	2,860	2,490	2,521
Unit cost of 2010 program before cost growth (millions of 2002 $)	45.5	62.8	60.2
Total cost of 2010 program before cost growth (billions of 2002 $)	80.1	21.3	20.5
Unit cost from 2010 SAR (millions of 2002 $)	83.3	110.3	
Total cost from 2010 SAR (billions of 2002 $)	146.8	75.0	

Given this new production profile of U.S. and foreign variants, total production cost for the U.S. variants would be $80.1 billion for the CTOL version, $21.3 billion for the CV, and $20.5 for the STOVL, for a total of $121.9 billion. The 2010 production cost estimate for the CTOL is $146.8 billion, which is a cost growth over the production-adjusted 2001 estimate of 83 percent. The December 2010 SAR gives only a combined DON production cost of $75.0 billion. This is a 79 percent increase over the combined production-adjusted CV and STOVL cost of $41.8 billion. We apply these cost-growth factors to the 2001 production cost estimates of $80.5 billion for the CTOL and $64.6 billion for the DON aircraft. Thus, our total production-adjusted 2010 cost for the JSF program is $262.3 billion, an 81 percent cost growth factor over the 2001 SAR value of $145.1 billion.

We use an analogous approach to assess JSF O&S cost growth. We first calculate what O&S costs would be at the 2010 fleet sizes, using the O&S cost relation shown above. The estimated U.S. fleet size fell from 2,852 to 2,443 between the 2001 and 2010 SARs. This leads to a slight increase in the 2001 SAR-based estimate of O&S costs per aircraft from $53.3 million to $53.6 million, because of the lost economies of scale. Thus, the estimated O&S costs based on the 2010 fleet size and the 2001 SAR would be

$131.0 billion, as opposed to the $151.9 figure in the 2001 SAR.[8] O&S costs in the 2010 SAR are $420.3 billion, a cost growth factor of 221 percent. We apply this cost-growth factor to the 2001 SAR estimate of $151.9 billion to derive $487.4 billion as our production-adjusted 2010 JSF O&S cost estimate after cost growth, using the 2001 fleet size. (The main reason for the difference in the $420 billion and $487 billion figures is the higher number of aircraft in the production-adjusted estimate, of course. The loss of economies of scale is comparatively minor.)

Single-Service Aircraft Estimates with Cost Growth

Now we turn to our estimate of what three single-service programs would cost after cost growth. For this calculation, we use F-22 cost growth at a comparable point in its program. There was no F-22 SAR at nine years past the program's MS B contract award. For acquisition, we used the average of the SARs just before and just after that point in time to estimate the cost growth in the program at nine years. In the program's December 1999 SAR, about 8.4 years past the August 1991 MS B contract award, the development estimate increased 27 percent and the procurement estimate, adjusted to the program's current quantity at that time of 333 production aircraft, increased 3 percent. In the program's next SAR, dated September 2001 (about 10.1 years past the program's MS B), the development estimate increased 33 percent over its MS B estimate. The procurement estimate, now adjusted to 297 planned production aircraft, increased 32 percent over its MS B estimate. Averaging the cost-growth estimates from the two SARs resulted in the 30 percent for development and the 17 percent for procurement we used for acquisition cost growth in our notional single-service programs.

The process for calculating O&S cost growth was somewhat different. An examination of F-22 SARs shows that, unlike acquisition costs, O&S estimates were not updated every year. F-22 O&S cost estimate updates closest to the nine year point past MS B were made at seven and 9.7 years past the milestone. We chose to use the 27.5 percent O&S cost growth estimated at 9.7 years, which appeared in the program's September 2001 SAR and is considerably higher than prior cost-growth estimates. Cost per flight hour was used to measure O&S cost growth, with no adjustment for the effects of the reduced aircraft procurement quantity planned—from 648 at MS B to 297 in September 2001. We made these analytical choices to be conservative in our results. The

---

[8] We note here that we are implicitly assuming that there are constant economies of scale in the CV and STOVL aircraft in the O&S cost categories that we judged were not affected by the joint fleet size (see Appendix D.) This is equivalent to an assumption that the CV and STOVL fleet reductions will be accomplished by reducing the number of squadrons, with the same number of aircraft in each squadron as in the 2001 plan. Our discussions with SPO personnel lead us to believe that this is the appropriate assumption.

sensitivity of using a 23.3 percent growth rate, which would include the fleet size adjustment, is shown below. We also show the sensitivity of using the F-22 O&S cost growth at seven years past MS B (December 1998 SAR), which was 12.4 percent. Both of these increase the cost savings of the projected single-service programs over the JSF, of course.

These cost-growth percentages—for development, procurement, and O&S—were applied to the combined baseline cost estimates of three single-service programs with the results shown in Table E.7. Note that, for the R&D calculation, we again assumed that the UK contribution to the STOVL would remain unchanged at $2.2 billion (it has remained at that level in the current program), so the cost to the United States grows slightly higher than the overall cost growth.

We also did an excursion, described in the main report, using F-22 O&S cost growth at the program's initial operational capability (IOC), which occurred at 14 years past MS B. These data were reported in the program's December 2005 SAR. This SAR gave the very first update to the program's estimated O&S costs since the 27.5 percent cost growth from September 2001. In other words, F-22 SARs reported the same 27.5 percent O&S cost growth from September 2001 through December 2004. The level of cost growth at IOC in December 2005 was 89.8 percent. Applying this F-22 O&S cost-growth rate at 14 years past MS B to the baseline O&S estimate for our three notional single-service fighters results in an estimate of O&S costs for single-service programs of $300.3 billion and a total of $579.6 billion.

Table E.8 summarizes all the results from this appendix.

Table E.7
Cost of Single-Service Programs After F-22 Cost Growth Factors Are Applied

	R&D	Production	O&S	Total
Estimate before cost growth (in billions of 2002 $)	51.8	180.5	158.2	390.6
F-22 cost-growth factor (%)	30 [a]	17	27.5	
Estimate after cost growth (in billions of 2002 $)	68.0	211.2	201.8	481.0

[a] The 30 percent is applied to the entire RDT&E estimate at MS B, which includes the $2.2 billion in foreign contributions. As a result, the dollar value of 30 percent cost growth is $16.2 billion, giving the total U.S. cost of $68 billion after cost growth.

Table E.8
Summary of Results: JSF and Single-Service Program Costs (at 2001 Production Levels)
With and Without Cost Growth (2002 $)

	JSF Before Cost Growth	Single-Service Programs Before Cost Growth	JSF After Cost Growth	Single-Service Programs After Cost Growth
R&D	30.2	51.8	46.2	68.0
Production	145.1	180.5	262.3	211.2
Acquisition	175.3	232.3	308.5	279.2
O&S	151.9	158.2	487.4	201.8
Total	327.3	390.6	795.9	481.0
Single-service O&S estimated using 14 years after MS B cost-growth factor				300.3
Total with single-service O&S estimated using 14 years after MS B cost-growth factor				579.6

Sensitivity Analysis

We now discuss a sensitivity analysis of our results. This is appropriate because a question arises as to how much confidence we put in our results. We cannot formally estimate a "confidence interval" in the statistical sense, because our results are not based on a formal estimation procedure using historical data—there are no historical experiments on which to base such a procedure. Therefore, we show the sensitivity of our results to what we have judged to be reasonable excursions on our input variables. In our overall professional judgment, we can have high confidence that the JSF program would not save significant amounts of LCC over what three single-service programs would have cost.

We begin with the sensitivity to the assumed characteristics of the "single-service programs after cost growth." As stated above, we used the acquisition cost-growth values of the F-22, 30 percent for development and 17 percent for procurement, for our single-service program projection. An interesting excursion is to use instead the averages from all the single-service programs in Appendixes B and C. These are 25 percent for development and 24 percent for procurement. This increases the total single-service cost, including O&S, from $481 billion to $487 billion; this compares with the JSF after cost growth of $796 billion. Thus, total savings of the projected single-service programs are reduced from 39.6 percent to 38.3 percent. The change in acquisition cost is more pronounced, of course. Single-service cost increases from $279 billion to $289 billion; this compares with the JSF after cost growth of $309 billion. Thus, total savings are reduced from 9.5 percent to 6.3 percent.

Another interesting excursion would be to determine the sensitivity of results to changing the assumption in the original JSF program—that RDT&E for three single-service programs would be 1.67 times JSF RDT&E—to either 1.5 or 2. Because RDT&E is a relatively small part of total cost, the total is not changed much, but the change in RDT&E savings is of interest. If the factor is 1.5 instead of 1.67, the increased RDT&E costs of the projected single-service programs are 32.0 percent instead of 47.2 percent. Here, we are implicitly assuming that there were fewer inherent economies of RDT&E in the joint program, and thus a single-service RDT&E program would have less of an RDT&E penalty. The increased cost of the overall single-service programs over that of the JSF after cost growth is 40.5 percent instead of 39.6 percent. If the factor is 2.0 instead of 1.67, the increased RDT&E costs of the projected single-service programs are 77.5 percent instead of 47.2 percent. Here, we are implicitly assuming that there were greater inherent economies of RDT&E in the joint program, and thus a single service RDT&E program would have a larger RDT&E penalty. The increased cost of the overall single-service RDT&E programs over that of the JSF after cost growth is 37.8 percent instead of 39.6 percent.

We also conducted two sensitivity analyses of our assumptions about F-22 O&S cost growth, which we applied to our projected single-service programs. The first was applying a fleet size factor to the 27.5 percent cost growth that was observed. As stated in the text, the planned size of the F-22 fleet also decreased over time, from 648 at MS B to 297 in September 2001. This 46 percent decrease should lead to a 3.4 percent increase in unit O&S costs, thus the cost growth is only 23.3 percent. Applied to the notional single-service fighters, this leads to 60.4 percent savings in O&S for the single-service programs versus the JSF, compared with 58.6 percent in the earlier results. Thus, it is not a significant factor. The second excursion entailed using the F-22 O&S cost growth at seven years past MS B (December 1998 SAR), which was 12.4 percent. This leads to 63.5 percent savings in O&S for the single-service programs versus the JSF, compared with 58.6 percent in the earlier results. Thus, it is not a significant factor.

Another potentially informative excursion is to examine what the sensitivity of the results would be to changing the assumption on commonality among the JSF variants. As discussed in the main report, the goal of DoD was to achieve 80 percent commonality among all three JSF variants, and we use that as the commonality among the three U.S. variants, as well as the commonality of the same variant between the U.S. and foreign aircraft.

How would changing this assumption affect outcomes? We examined 60 percent and 90 percent commonality excursions. As one would expect, the higher the degree of commonality, the less is the cost advantage of the projected single-service program over the JSF. In the 60 percent case, the procurement cost savings of the projected single-service program are 22.1 percent compared with 19.5 percent in the base case, and the total program savings are 40.5 percent compared with 39.6 percent in the base case. Since commonality is relatively less important here, the single-service programs are relatively less costly. In the 90 percent case, the procurement cost savings of the projected single-service program are 18.3 percent compared with 19.5 percent in the base case, and the total program savings are 39.2 percent compared with 39.6 percent in the base case. Since commonality is relatively more important here, the single-service programs are relatively more costly.

Appendix F: Alternative Procurement Methodology and Results for Comparing JSF Costs with Those of Three Notional Single-Service Fighters

As shown in Appendix E, the primary methodology for estimating the procurement costs of the three notional single-service fighter programs is derived from F-35 program estimates. This appendix provides an alternative assessment of the procurement cost of three independent single-service programs that build the three JSF variants.

The alternative approach changes only the estimates for procurement of three notional single-service programs; the actual JSF procurement baseline and F-35 (JSF) procurement estimate at nine years past MS B are the same as in the primary methodology. Development and O&S estimates are also the same as in the primary methodology.[1]

The results of the alternative methodology are similar to those of the primary methodology discussed in Appendix E. In general, the alternative approach shows slightly higher costs for the notional three single-service programs. The approach selects a unit cost value using credible data and then adjusts that value to the three, individual JSF variants.

Defining the Relationship Between Average Unit Procurement Cost (APUC) and Weight

Figure F.1 shows the most recent update of the internal RAND assessment "tool" for aircraft APUC.[2] The figure shows cumulative average APUC of the first 100 units for 13 unclassified military aircraft programs, normalized to base-year 2012 dollars. For each program, the aircraft's APUC is plotted against its empty weight. Only aircraft built from a "cold" production line are included. The center line on the chart was determined by regression analysis of the 13 data points. The resulting slope is 159.2 percent, indicating that if any particular aircraft design doubled in weight, its cost would increase by

---

[1] The development estimates are derived from detailed information from the JSF program office as explained above. No alternative method to those estimates, or those for the O&S estimates, is made.

[2] The graphic was derived from December 2009 and earlier SAR data. Similar tools are used by cost analysis professionals at the Air Force Cost Analysis Center, Naval Center for Cost Analysis, OSD's Cost Analysis Improvement Group, and the major aerospace manufacturing corporations.

59.2 percent.[3] The slope of the lines in Figure F.1 are equivalent to a 79.6 percent (159.2 percent ÷ 2) cost per pound relationship with the doubling of weight. This is consistent with the generally held view that aircraft production cost per pound versus pounds scales on a slope in the 80 percent to 85 percent range.

Figure F.1
Normalized Historical Military Aircraft APUC

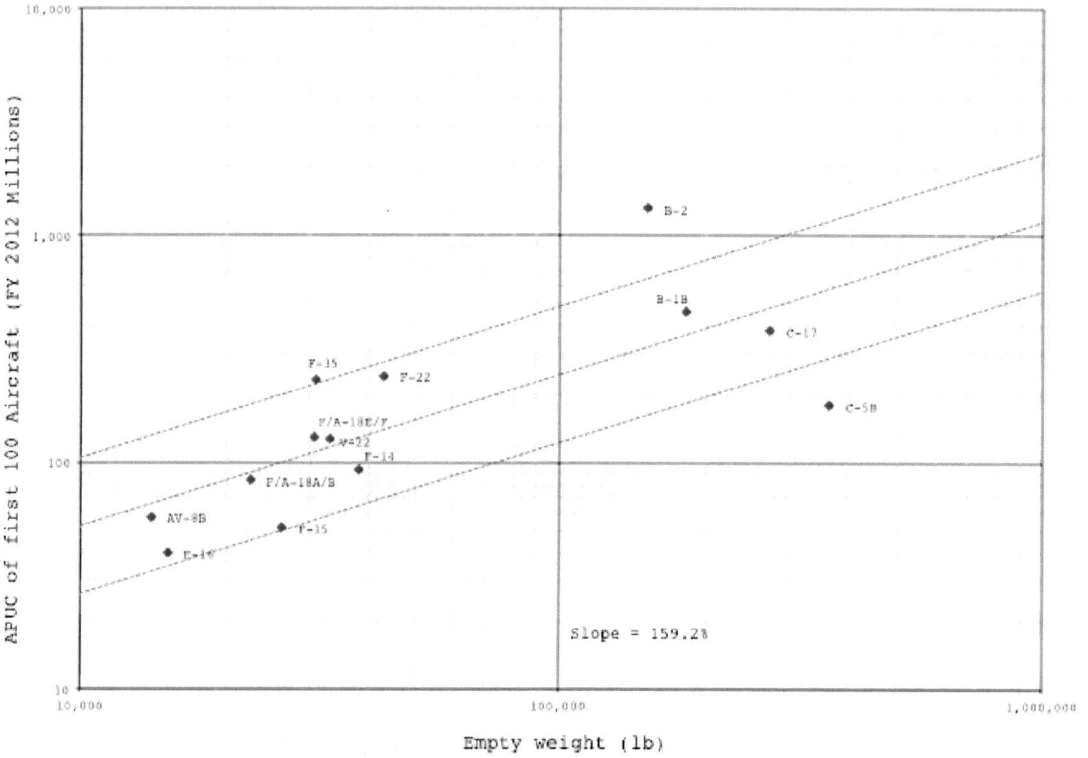

As drawn in Figure F.1, the upper and lower lines differ from the center line by a factor of two. The upper line is twice as high as the center line, and the lower line is half as high. This spacing looks reasonable relative to the distribution of the plot points.[4]

---

[3] For example, the unit cost of a 20,000 lb EW aircraft would be 159.2 percent that of the cost of a 10,000 lb version of the same aircraft. Cost analysts most often develop such curves using cost per pound versus weight. Of course, other data sets may result in different slopes. In this form, the lines are called ARCO curves, following their use to support aircraft production during World War II. For more on ARCO curves see Joseph P. Large, Hugh G. Campbell, and D. Cates, *Parametric Equations for Estimating Aircraft Airframe Costs*, Santa Monica, Calif.: RAND Corporation, R-1693-1-PAE, 1975.

[4] The standard error of the estimate for the regression is 0.6878 (the regression is done in logarithms), which is equivalent to a factor of 1.989; this compares with the factor of 2.0 used in the plot.

The rationale behind the three lines on the figure may not be immediately intuitive. Considering which aircraft fall near the upper, middle, and lower lines and knowing something about their characteristics, history, and difficulties of working with data of varying vintages, each line can reasonably be interpreted as representing a different complexity of design. The lowest represents older aircraft designs that included relatively less-sophisticated avionics and software and no stealth characteristics. Three aircraft fall close to or below this line—the F-16, F-15, and C-5B. The middle line represents newer designs, which are more complex and include increased amounts of software, but still no stealth. The highest line represents the most modern designs, which are the most complex and make the most intensive use of sophisticated avionics and software. These aircraft make the most comprehensive use of the latest stealth technology available at the time of their development. Three aircraft fall close to or above the highest line—the F-35, F-22, and B-2. Observing that the F-22 and F-35 are both near the upper line is additional support for using F-22 program costs as the basis of the estimates developed below.

Although stealth and systems complexity help explain the position of the reference lines relative to the scatter of points, other factors help explain the vertical position of the programs relative to the lines. The values used to create Figure F.1 were determined by fitting cost-improvement curves to the early procurement lots for each program. The lots included were selected by examining plots of the lot unit costs versus plot points to identify breaks in the curves that would represent significant model changes or other program perturbations. If possible, lots were included up to a cumulative procurement quantity of at least 100 units. The progress curve analyses did not include foreign sales or variations in production rate. Both of these factors can change the values shown in the figure.

Production rates influence total cost via the allocation of fixed overhead costs—the higher the production rate, the more units over which to spread fixed costs. Production rates vary over the course of a given program and they vary between programs. Additional research is necessary to determine the efficacy of making adjustments between programs, but we note that, for the aircraft in Figure F.1, the B-2 has the lowest production rate and the F-16 has the highest production rate. For the data used to determine the APUC values at 100 units, the average annual rate was approximately 36. Using a 90 percent rate normalization slope, the B-2 APUC would decrease about 35 percent and the F-16 APUC would increase by about 25 percent. The other points would shift by smaller amounts—the AV-8B, F-14, F/A-18A/B, F/A-18E/F, and F-35 would move by less than 3 percent.

Calculating Alternative MS B Baseline Estimates for the Three Notional Single-Service Fighters the Size of the JSF Variants

We begin with the APUC value estimated at the F-22 program's MS B.[5] We adjust that estimate to the weight of the three JSF variants using the relationship between aircraft empty weight and cost as defined in Figure F.1. By keying our estimates to those of the F-22, we embody in them costs for a high performance, state-of-the art fighter acquisition program.[6] To perform the EW-to-cost adjustments, we use the relationship defined in Figure F.1.

Because the weight versus cost relationship used is at the bottom end of the accepted range for this relationship (79.6 percent versus the commonly accepted values of 80 percent to 85 percent), and because all three JSF variants are smaller than the F-22, the MS B estimates for the three notional single-service programs building the JSF variants are on the high end of the range defined by the slope of the weight versus cost curve, giving the maximum possible advantage to the joint program. Using these factors, procurement costs of the three single-service aircraft are estimated as individual programs. The estimates and relevant data for each are shown in Table F.1.

Table F.1
F-35 Variant Procurement MS B Estimates from Alternative Methodology Scaled from F-22 Costs

	F-22	CTOL	STOVL	CV
Aircraft EW (lb)	42,000	29,300	32,300	34,800
Theoretical first unit cost (in millions of BY 2002 $)	228.3	179.3	191.4	201.2
Cost-quantity improvement curve (%)	90.24	90.24	90.24	90.24
MS B APUC of first 100 aircraft (in millions of BY 2002 $)	115.4	90.6	96.8	101.7
MS B procurement quantity	648	1,763	609	480
MS B average procurement unit cost for the MS B quantity (in millions of BY 2002 $)	87.5	59.3	72.1	80.6
Alternative methodology MS B estimate (in millions of BY 2002 $)	56,700	104,500	43,900	38,700

---

[5] This is not the data point shown on Figure F.1.

[6] The F-22 was a relatively high-risk development program because it was the first supersonic/super cruise LO fighter ever developed. It was also the first fighter with a fully integrated avionics system. From the very beginning of RDT&E, the F-22 experienced significant cost growth and was a target of extensive criticism from Congress, the Government Accountability Office, and the press.

The theoretical first-unit cost in each notional single-service program is derived from the 100-unit APUC, which was derived from using the direct relationship between cost and weight for each aircraft. The 90.24 percent cost-quantity improvement curve is then applied to each of the single-service programs, along with the estimated first-unit cost in each program and the number of units (U.S. plus foreign) to determine the APUC for the production run of each aircraft. That figure is multiplied by the number of U.S.-purchased aircraft to determine the cost to the United States for its planned procurement. In all cases, MS B quantities apply, including 150 U.K. STOVL aircraft that are assumed to be 80 percent common with the U.S. version.

Summing the procurement estimates for the three single-service aircraft programs at the bottom of Table F.1 gives a baseline cost estimate in aggregate of $187.1 billion ($104.5 billion CTOL + $43.9 billion STOVL + $38.7 billion CV). Using the primary methodology, as shown in Tables E.4, E.6, and E.7, the estimate is $180.5 billion, or $6.6 billion (3.6 percent) less.

In Figure F.2 we compare acquisition estimates for MS B baseline and nine years past that baseline using the alternative procurement estimating methodology for procurement costs. This figure is similar to Figure 3.2 in the main report, which shows the same comparison using the primary methodology.

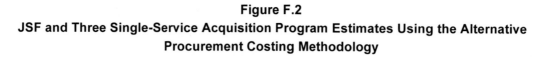

Figure F.2
JSF and Three Single-Service Acquisition Program Estimates Using the Alternative Procurement Costing Methodology

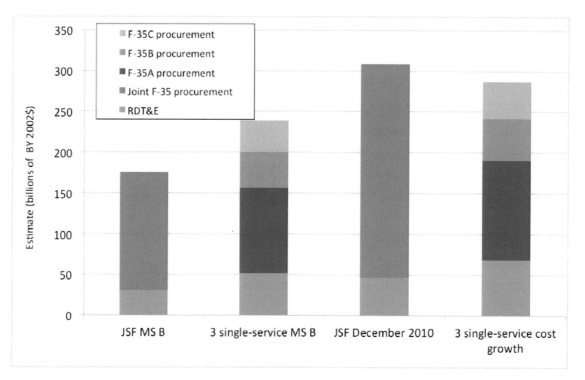

The leftmost stacked bar in Figure F.2 shows the JSF MS B acquisition estimate of $175.3 billion. This includes $30.2 billion in development and $145.1 billion in procurement for 2,852 aircraft. The stacked bar to its right shows our estimate at MS B of $238.9 billion for the DoD acquisition cost for the three single-service fighter programs using the alternative procurement methodology. This estimate includes $51.8 billion in development and $187.1 billion in procurement. Comparing the joint with the single-service acquisition approach shows that the joint program is estimated to cost $63.5 billion ($238.9 billion – $175.3 billion, or 27 percent) less than the three single-service fighter programs at MS B. This difference using the primary methodology is $57.0 billion ($232.3 billion – $175.3 billion), or 24.5 percent.

The MS B estimate of $51.8 billion in development for the three single-service fighter programs is assumed to be equal to 1.667 times the JSF estimate.[7] Both the JSF estimate and three single-service program MS B estimates assume a fixed $2.2 billion foreign contribution for the STOVL aircraft.

The MS B production estimate for the three single-service programs assumes the observed F-22 cost-quantity improvement curve of 90.24 percent, first-unit costs of each

---

[7] This ratio is based on JPO and contractor claims.

aircraft based on JSF-variant weight estimates, an ARCO (cost versus weight) curve of 159.2 percent, and 80 percent commonality between U.S. and UK STOVL aircraft.

The rightmost two stacked bars in Figure F.2 show the estimates at nine years past MS B. The December 2010 SAR acquisition cost for the JSF program, adjusted to 2001 quantities and foreign participation, is $308.5 billion. The estimate includes $46.2 billion for development, adjusted so that the additional $2.2 billion foreign contribution since 2001 is not charged to the DoD, plus $262.3 billion for production, calculated using the following adjustments:

- adjusting for quantity changes (rebaselining to the original 480 CV and 609 STOVL U.S. buys versus the current 340 and 340 U.S. buys)
- adjusting assumptions for foreign sales to the baseline assumption of zero CV and 150 STOVL foreign buys versus the 2010 SAR assumption of 730 CTOL and STOVL foreign aircraft
- incorporating the assumption of 80 percent commonality between U.S. and foreign versions.

With these adjustments, and using the alternative methodology, JSF cost growth in acquisition is 76 percent from the program's MS B in late 2001 through its December 2010 SAR estimate.

The same F-22 cost-growth factors at nine years past MS B used in the primary methodology were applied to the three single-service fighter programs' estimates derived from the alternative procurement methodology. The results are shown in the rightmost stacked bar of Figure F.2. The result is $286.9 billion total acquisition for the three programs: $68.0 billion for development, $122.2 billion for 1,743 CTOL aircraft, $51.4 billion for 608 STOVL aircraft, and $45.3 billion for 480 CV aircraft. The F-22 cost-growth factors applied were 30 percent RDT&E cost growth and 17 percent production cost growth. No quantity adjustments are required to calculate this estimate because F-22 cost-growth factors are applied directly to the MS B baselines for the programs.

At nine years past MS B, after including cost growth, the three single-service fighter programs acquisition estimate is $21.6 billion less than that for the JSF program ($308.5 billion from the JSF December 2010 SAR versus $286.9 billion for three single-service), representing a 7 percent saving. This compares with 9.5 percent in the primary methodology in which the three single-service fighter programs acquisition estimate is estimated at $29.3 billion less than that for the joint program ($308.5 billion from the JSF December 2010 SAR-- $279.2 billion for three single-service).

Figure F.3 adds O&S costs to the acquisition costs shown in Figure F.2 and is comparable to Figure 3.4 in the main report, which shows the comparison using the primary methodology. F-22 O&S cost-growth percentages from both 9.7 and 14 years past that program's MS B are used.

At MS B, the O&S cost estimate for three single-service fighter programs is $158.2 billion. This compares with $151.9 billion in the joint program at MS B. These costs are shown as the top bars in the figure's two left-hand stacked bars. The 4 percent savings in the joint program comes from a larger percentage savings expected in the nearly two-thirds of O&S costs per aircraft that benefit from economies of scale. The remaining one-third of O&S costs do not experience economies of scale.

Figure F.3
JSF and Three Single-Service LCC Program Estimates Using the Alternative Procurement Costing Methodology, Billions of BY 2002 Dollars

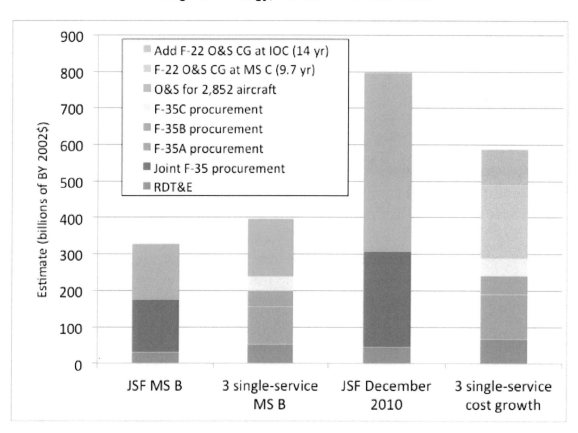

Adding the JSF MS B program baseline estimates for acquisition and O&S gives a total LCC estimate of $327.3 billion ($175.3 billion acquisition + $151.9 billion O&S). This is the left-most stacked bar. The corresponding estimate for the three single-service programs is $397.1 billion ($238.9 billion acquisition + $158.2 billion O&S). This is the stacked bar second from the left. As a result, at MS B, the JSF program shows a $69.8 billion savings ($397.1 billion – $327.3 billion), or 17.6 percent, over the estimated cost of the three single-service fighter programs. This compares with $63.2 billion ($390.6 billion – $327.3 billion), or 16.2 percent, JSF savings estimated using the primary methodology.

The DoD JSF O&S cost estimate at MS B was $151.9 billion. As of December 2010, the F-35 (JSF) SAR estimated that figure would increase 221 percent to $487.4 billion, after adjusting to the MS B quantity of 2,852. This is the top bar of the second from the right stacked bar. Using the F-22 O&S cost-growth factor of 27.5 percent at MS C (9.7 years past MS B), the three single-service fighter programs' O&S cost estimates increase from $158.2 billion to $201.8 billion. This is the second from the top bar in the right-most stacked bar. When using the F-22 O&S cost-growth factor of 89.8 percent at IOC (14 years past MS B), we see that the three single-service fighter programs' O&S costs increase another $98.5 billion (the top bar in the right-most stacked bar), for a total of $300.3 billion.

The December 2010 DoD F-35 (JSF) LCC , adjusted to 2001 quantities and foreign participation, is $795.9 billion ($308.5 billion acquisition + $487.4 billion O&S). This is shown in the tallest stacked bar in Figure F.3.

With F-22 cost-growth factors for acquisition and O&S at MC C (9.7 years past MS B), the LCC for the three single-service fighter programs are $488.7 billion ($286.9 billion acquisition + $201.8 billion O&S). This is $307.2 billion, or 38.6 percent less ([$488.7 billion/$795.9 billion] – 1 = –0.386) than the JSF joint program estimate. Put another way, the JSF LCC estimate is 63 percent more ($795.9 billion/$488.7 billion = 1.629) than the three single-service programs at nine years past MS B. Using the primary methodology, the JSF is 65 percent more ($795.9 billion/$481.0 billion = 1.6547) than the three single-service programs at this point.

With F-22 cost-growth factors on acquisition and O&S at IOC (14 years past MS B), the LCC for three single-service fighter programs is $587.2 billion ($286.9 billion acquisition + $300.3 billion O&S). This is the right-most stacked bar. This is $208.7 billion, or 26 percent less ([$587.2 billion/$795.9 billion] – 1 = –0.262) than the JSF LCC. Put another way, the JSF LCC estimate is 36 percent more ($795.9 billion/$587.2 billion = 1.355) than the three single-service programs. Using the primary methodology, the JSF is 37 percent more ($795.9 billion/$579.6 billion = 1.373) than the three single-service programs.

Bibliography

Aghion, Philippe, "Empirical Estimates of the Relationship Between Product Market Competition and Innovation," in Jean-Philippe Touffut, ed., *Institutions, Innovation and Growth: Selected Economic Papers*, Cheltenham, UK: Elgar, Saint-Gobain Centre for Economic Studies Series, 2003, pp. 142–169.

Aghion, Philippe, Nick Bloom, Richard Blundell, Rachel Griffith, and Peter Howitt, "Competition and Innovation: An Inverted-U Relationship," *Quarterly Journal of Economics*, Vol. 120, No. 2, May 2005, pp. 701–728.

Ahn, Sanghoon, *Competition, Innovation and Productivity Growth: A Review of Theory and Evidence*, Organisation for Economic Co-Operation and Development, Economics Department Working Paper 317, January 17, 2002.

Anderson, Fred, *Northrop: An Aeronautical History*, Los Angeles, Calif.: Northrop Corporation, 1976.

Anderton, David A., *Republic F-105 Thunderchief*, London: Osprey Publishing, 1983.

———, *North American F-100 Super Sabre*, London: Osprey Publishing, 1987.

Angelmar, Reinhard, "Market Structure and Research Intensity in High-Technological-Opportunity Industries," *Journal of Industrial Economics*, Vol. 34, No. 1, September 1985, pp. 69–79.

Arena, Mark V., Obaid Younossi, Kevin Brancato, Irv Blickstein, and Clifford A. Grammich, *Why Has the Cost of Fixed-Wing Aircraft Risen? A Macroscopic Examination of the Trends in U.S. Military Aircraft Costs over the Past Several Decades*, Santa Monica, Calif.: RAND Corporation, MG-696-NAVY/AF, 2008. As of February 29, 2012:
http://www.rand.org/pubs/monographs/MG696.html

Armitage, Michael J., *Unmanned Aircraft*, London: Brassey's Defense Publishers, 1988.

Aronstein, David C., and Michael J. Hirschberg, "An Overview of the US/UK ASTOVL Programs, 1983–1994," *International Powered Lift Conference, AHS International*, Paper IPLC-2000-00039, 2000.

Aronstein, David C., Michael J. Hirschberg, and Albert C. Piccirillo, *Advanced Tactical Fighter to F-22 Raptor: Origins of the 21st Century Air Dominance Fighter*, Reston, Va.: American Institute of Aeronautics and Astronautics, 1998.

Art, Robert J., *The TFX Decision: McNamara and the Military*, Boston: Little, Brown, 1968.

Baker, David, "From ATF to Lightning II: A Bolt in Anger—Part One: Design Options and the YF-23A," *Air International*, Vol. 47, No. 6, December 1994.

Barkey, H. D., *Evolution of the F-4 Phantom*, presented at the Technical Program Management Seminar of the American Institute of Industrial Engineers, Washington, D.C., April 19–21, 1971.

Bevilaqua, Paul M., "Joint Strike Fighter Dual-Cycle Propulsion System," *Journal of Propulsion and Power*, Vol. 21, No. 5, September–October 2005.

———, "Genesis of the F-35 Joint Strike Fighter," *Journal of Aircraft*, Vol. 46, No. 6, November–December 2009.

Birkler, John, Edmund Dews, and Joseph P. Large, *Issues Associated with Second-Source Procurement Decisions*, Santa Monica, Calif.: RAND Corporation, R-3996-RC, December 1990. As of February 29, 2012:
http://www.rand.org/pubs/reports/R3996.html

Birkler, John, John C. Graser, Mark V. Arena, Cynthia R. Cook, Gordon T. Lee, Mark A. Lorell, Giles K. Smith, Fred Timson, Obaid Younossi, and Jon Grossman, *Assessing Competitive Strategies for the Joint Strike Fighter: Opportunities and Options*, Santa Monica, Calif.: RAND Corporation, MR-1362-OSD/JSF, 2001. As of February 29, 2012:
http://www.rand.org/pubs/monograph_reports/MR1362.html

Birkler, John, Anthony G. Bower, Jeffrey A. Drezner, Gordon T. Lee, Mark A. Lorell, and Giles K. Smith, *Competition and Innovation in the U.S. Fixed-Wing Military Aircraft Industry*, Santa Monica, Calif.: RAND Corporation, MR-1656-OSD, 2003. As of February 29, 2012:
http://www.rand.org/pubs/monograph_reports/MR1656.html

Birkler, John, Paul Bracken, Gordon T. Lee, Mark A. Lorell, Soumen Saha, and Shane Tierney, *Keeping a Competitive U.S. Military Aircraft Industry Aloft: Findings from an Analysis of the Industrial Base*, Santa Monica, Calif.: RAND Corporation, MG-1133-OSD, 2011. As of February 29, 2012:
http://www.rand.org/pubs/monographs/MG1133.html

Bolkcom, Christopher C., *Tactical Aircraft Modernization: Issues for Congress*, Washington, D.C.: Congressional Research Service, IB92115, July 3, 2002.

———, *Proposed Termination of Joint Strike Fighter (JSF) F136 Alternate Engine*, Washington, D.C.: Congressional Research Service, RL33390, April 13, 2006.

Bolton, Joseph G., Robert S. Leonard, Mark V. Arena, Obaid Younossi, and Jerry M. Sollinger, *Sources of Weapon System Cost Growth: Analysis of 35 Major Defense Acquisition Programs*, Santa Monica, Calif.: RAND Corporation, MG-670-AF, 2008. As of April 9, 2012:
http://www.rand.org/pubs/monographs/MG670.html

Braybrook, Roy, "F-14 and F-15: The New Wave of Warplanes," *Air Enthusiast*, March 1972.

Bright, Charles D., *The Jet Makers: The Aerospace Industry from 1945 to 1972*, Lawrence, Kan.: Regents Press of Kansas, 1978.

Burton, James, *The Pentagon Wars: Reformers Challenge the Old Guard*, Annapolis, Md.: Naval Institute Press, 1993.

Congressional Budget Office, *Balance and Affordability of the Fighter and Attack Aircraft Fleets of the Department of Defense*, Washington, D.C., April 1991.

Coulam, Robert F., *Illusions of Choice: The F-111 and the Problems of Weapons Acquisition Reform*, New York: Princeton University Press, 1977.

Davis, Charles R. (Lt Gen, USAF), program executive officer, F-35 Program Office, "JSF Production," briefing presented at the Aviation Week Aerospace and Defense Finance Conference, New York, November 2008.

Delusach, Al, "Long Preparation Paid Off for McDonnell on the F-15," *St. Louis Post-Dispatch*, January 11–12, 1970.

Department of Defense Instruction (DoDI) 5000.02, *Operation of the Defense Acquisition System*, December 8, 2008. As of June 27, 2012:
http://www.dtic.mil/whs/directives/corres/pdf/500002p.pdf

Donald, David, ed., *The Complete Encyclopedia of World Aircraft*, New York: Barnes Noble Books, 1999.

Dorr, Robert F., *Lockheed F-117 Nighthawk*, London: Aerospace Publishing, 1995.

Drezner, Jeffrey A., Giles K. Smith, Lucille E. Horgan, J. Curt Rogers, and Rachel Schmidt, *Maintaining Future Military Aircraft Design Capability*, Santa Monica, Calif.: RAND Corporation, R-4199-AF, 1992. As of February 29, 2012:
http://www.rand.org/pubs/reports/R4199.html

Ethell, Jeffrey L., *F-15 Eagle*, London: Ian Allan, 1981.

F-22 Program Office, Col. Sean M Frisbee, 478 AESG/CC, *Selected Acquisition Report: F-22*, Wright-Patterson Air Force Base, Ohio, December 31, 2009.

———, ASC/WWU, *Selected Acquisition Report: F-22*, Wright-Patterson Air Force Base, Ohio, December 31, 2010.

F-35 Joint Strike Fighter Program Office, Brig Gen John Hudson, JSFPO, *Selected Acquisition Report: Joint Strike Fighter*, Arlington, Va., December 31, 2001.

F-35 Lightning II Program Office, VADM David Venlet, *Selected Acquisition Report: F-35*, Arlington, Va., December 31, 2010.

Farrell, Joseph, Richard J. Gilbert, and Michael L. Katz, "Market Structure, Organizational Structure, and R&D Diversity," in Richard Arnott, ed., *Economics for an Imperfect World: Essays in Honor of Joseph E. Stiglitz*, Cambridge, Mass.: MIT Press, 2003, pp. 195–220.

Francillon, René, *McDonnell Douglas F-15A/B*, Arlington, Texas: Aerofax, 1984.

———, *McDonnell Douglas Aircraft Since 1920: Volume II*, Annapolis, Md.: Naval Institute Press, 1990.

Gansler, Jacques S., William Lucyshyn, and Michael Arendt, *Competition in Defense Acquisitions*, College Park, Md.: University of Maryland Center for Public Policy and Private Enterprise, School of Public Policy, revised February 2009. As of February 29, 2012:
http://www.dtic.mil/dtic/tr/fulltext/u2/a529406.pdf

Gentry, Jerauld R., *Evolution of the F-16 Multinational Fighter*, Washington, D.C.: Industrial College of the Armed Forces, 1976.

Geroski, Paul A., "Entry, Innovation and Productivity Growth," *Review of Economics and Statistics*, Vol. 71, No. 4, November 1989, pp. 572–578.

Gertler, Jeremiah, *F-35 Joint Strike Fighter (JSF) Program: Background and Issues for Congress*, Washington, D.C.: Congressional Research Service, RL30563, April 26, 2011.

Gething, Michael J., *F-15*, New York: Arco Publishing, 1983.

Gilbert, Richard J., "Competition and Innovation," *Journal of Industrial Organization Education*, Vol. 1, No. 1, 2006.

Godfrey, David W. H., "Dogfighter Supreme: The Tomcat," *Air Enthusiast International*, January 1977.

Goodwin, Jacob, *Brotherhood of Arms: General Dynamics and the Business of Defending America*, New York: Times Books, 1985.

Green, William, and Gordon Swanborough, *The Complete Book of Fighters*, New York: Smithmark, 1994.

Gunston, Bill, *Attack Aircraft of the West*, New York: Scribner, 1974.

———, *The Encyclopedia of the World's Combat Aircraft*, New York: Chartwell, 1978.

———, *The Encyclopedia of World Air Power*, New York: Crescent Books, 1980.

———, *F-111*, New York: Arco Publishing, 1983.

———, *Modern Fighting Aircraft*, New York: Military Press, 1984.

———, *F/A-18 Hornet*, London: Ian Allen, 1985.

Hale, Robert F., assistant director, National Security Division, Congressional Budget Office, testimony before the Subcommittee on Conventional Forces and Alliance Defense, Committee on Armed Services, U.S. Senate, Washington, D.C., April 22, 1991. As of April 9, 2012:
http://handle.dtic.mil/100.2/ADA529821

Hallion, Richard P., *The Evolution of Commonality in Fighter and Attack Aircraft Development and Usage*, Edwards Air Force Base, Calif.: Air Force Flight Test Center History Office, November 1985.

———, "A Troubling Past: Air Force Fighter Acquisition Since 1945," *Air Power Journal*, Winter 1990, pp. 4–23. As of February 29, 2012:
http://www.airpower.au.af.mil/airchronicles/apj/apj90/win90/1win90.htm

Harman, Alvin J., *Analysis of Aircraft Development*, Santa Monica, Calif.: RAND Corporation, P-4976, 1973. As of February 29, 2012:
http://www.rand.org/pubs/papers/P4976.html

Harman, Alvin J., and S. Henrichsen, *A Methodology for Cost Factor Comparison and Prediction*, Santa Monica, Calif.: RAND Corporation, RM-6269-ARPA, August 1970. As of February 29, 2012:
http://www.rand.org/pubs/research_memoranda/RM6269.html

Head, Richard G., *Decision Making on the A-7 Attack Aircraft Program*, Syracuse, N.Y.: Syracuse University, Ph.D. thesis, December 18, 1970.

———, "Doctrinal Innovation and the A-7 Attack Aircraft Decision," in Richard G. Head and Ervin J. Rokke, eds., *American Defense Policy*, 3rd ed., Baltimore, Md.: Johns Hopkins University Press, 1973.

59

Hehs, Eric, "F-22 Design Evolution, Part I," *Code One Magazine*, April 1998a. As of April 9, 2012:
http://www.codeonemagazine.com/article.html?item_id=40

———, "F-22 Design Evolution, Part II," *Code One Magazine*, October 1998b. As of April 9, 2012:
http://www.codeonemagazine.com/article.html?item_id=41

———, "F-16 Evolution," *Code One Magazine*, September 2008. As of April 9, 2012:
http://www.codeonemagazine.com/article.html?item_id=23

Held, Thomas, Bruce Newsome, and Matthew W. Lewis, *Commonality in Military Equipment: A Framework to Improve Acquisition Decisions*, Santa Monica, Calif.: RAND Corporation, MG-719-A, 2008. As of April 13, 2012:
http://www.rand.org/pubs/monographs/MG719.html

Joint STARS Program Office, Col. Charles E. Franklin, *Selected Acquisition Report: Joint STARS*, Hanscom Air Force Base, MA, December 31, 1985.

Johnson, Leland, *The Century Series Fighters: A Study in Research and Development*, Santa Monica, Calif.: RAND Corporation, RM-2549, May 1960. As of February 29, 2012:
http://www.rand.org/pubs/research_memoranda/RM2549.html

Jones, Lloyd S., *U.S. Fighters: Army-Air Force, 1920–1980s*, Fallbrook, Calif.: Aero Publishers, 1975.

———, *U.S. Naval Fighters*, Fallbrook, Calif.: Aero Publishers, 1977.

Kelly, Orr, *Hornet: The Inside Story of the F/A-18*, Shrewsbury, UK: Airlife, 1991.

———, *Lockheed Skunk Works*, Osceola, Wis.: Motorbooks International, 1992.

Klein, Burton H., William Meckling, and Emmanuel G. Mesthene, *Military Research and Development Policies*, Santa Monica, Calif.: RAND Corporation, R-333, December 1958. As of February 29, 2012:
http://www.rand.org/pubs/reports/R333.html

Knaack, Marcelle Size, *Post–World War II Fighters: 1945–1973*, Washington, D.C.: Office of Air Force History, 1978.

———, *Post–World War II Bombers: 1945–1973*, Washington, D.C.: Office of Air Force History, 1988.

Large, Joseph P., Hugh G. Campbell, and D. Cates, *Parametric Equations for Estimating Aircraft Airframe Costs,* Santa Monica, Calif.: RAND Corporation, R-1693-1-PAE, 1975. As Of March 25, 2013:
http://www.rand.org/pubs/reports/R1693-1.html

Li, Allen, *Tactical Aircraft: Status of the F/A-22 and Joint Strike Fighter Programs*, Testimony Before the Subcommittee on Tactical Air and Land Forces, Committee on Armed Services, House of Representatives, Washington, D.C.: U.S. General Accounting Office, GAO-04-597T, March 24, 2004. As of February 29, 2012:
http://purl.access.gpo.gov/GPO/LPS48489

Lorell, Mark, *The U.S. Combat Aircraft Industry, 1909–2000: Structure, Competition, Innovation*, Santa Monica, Calif.: RAND Corporation, MR-1696-OSD, 2003. As of February 29, 2012:
http://www.rand.org/pubs/monograph_reports/MR1696.html

Lorell, Mark A., and Donna K. Hoffman, *The Use of Prototypes in Selected Foreign Fighter Aircraft Development Programs: Rafale, EAP, Lavi, and Gripen*, Santa Monica, Calif.: RAND Corporation, R-3687-P&L, September 1989. As of February 29, 2012:
http://www.rand.org/pubs/reports/R3687.html

Lorell, Mark A., and Hugh P. Levaux, *The Cutting Edge: A Half Century of U.S. Fighter Aircraft R&D*, Santa Monica, Calif.: RAND Corporation, MR-939-AF, 1998. As of February 29, 2012:
http://www.rand.org/pubs/monograph_reports/MR939.html

Lorell, Mark A., Michael Kennedy, Robert S. Leonard, Ken Munson, Shmuel Abramzon, David L. An, and Robert A. Guffey, *Do Joint Fighter Programs Save Money?* Santa Monica, Calif.: RAND Corporation, MG-1225-AF, 2013. As of late 2013:
http://www.rand.org/pubs/monographs/MG1225.html

Lorell, Mark A., Julia F. Lowell, Michael Kennedy, and Hugh P. Levaux, *Cheaper, Faster, Better? Commercial Approaches to Weapons Acquisition*, Santa Monica, Calif.: RAND Corporation, MR-1147-AF, 2000. As of February 29, 2012:
http://www.rand.org/pubs/monograph_reports/MR1147.html

Lorell, Mark A., with Alison Saunders and Hugh P. Levaux, *Bomber R&D Since 1945: The Role of Experience*, Santa Monica, Calif.: RAND Corporation, MR-670-AF, 1995. As of June 27, 2012:
http://www.rand.org/pubs/monograph_reports/MR670.html

Lynch, David J., "How the Skunk Works Fielded Stealth," *Air Force Magazine*, Vol. 75, No. 11, November 1992. As of April 9, 2012:
http://www.airforce-magazine.com/MagazineArchive/Pages/1992/November%201992/1192stealth.aspx

Martin, Thomas, and Rachel Schmidt, *A Case Study of the F-20 Tigershark*, Santa Monica, Calif.: RAND Corporation, P-7495-RGS, June 1987. As of February 29, 2012:
http://www.rand.org/pubs/papers/P7495.html

Mason, Francis K., *Phantom: A Legend in Its Own Time*, Osceola, Wis.: Motorbooks International, 1983.

Mayer, Kenneth R., *The Political Economy of Defense Contracting*, New Haven, Conn.: Yale University Press, 1991.

Mendenhall, Charles A., *Delta Wings: Convair's High-Speed Planes of the Fifties and Sixties*, Osceola, Wis.: Motorbooks International, 1983.

Miller, Jay, *General Dynamics F-16 Fighting Falcon*, Arlington, Texas: Aerofax, 1982.

———, *Lockheed SR-71 (A/12YF-12/D-21)*, Arlington, Texas: Aerofax, 1983.

Morrocco, John D., "Northrop Sees ASTOVL as Inroad to JAST," *Aviation Week and Space Technology*, Vol. 141, No. 5, August 1, 1994.

Naval Air Systems Command, Col. J. A. Creech, USMC, *Selected Acquisition Report: V-22 (JVX)*, Naval Air Systems Command, Washington D.C., December 31, 1985.

Neufeld, Jacob, "The F-15 Eagle: Origins and Development, 1964–1972," *Air Power History*, Vol. 48, No. 1, Spring 2001.

Newsome, Bruce, Matthew W. Lewis, and Thomas Held, *Speaking with a Commonality Language: A Lexicon for System and Component Development*, Santa Monica, Calif.: RAND Corporation, TR-481-A, 2007. As of April 13, 2012:
http://www.rand.org/pubs/technical_reports/TR481.html

Office of the Under Secretary of Defense for Acquisition, *1983 Defense Science Board Summer Study Briefing Report for Joint Service Acquisition Programs*, Washington, D.C., August 1–12, 1983. As of April 9, 2012:
http://www.dtic.mil/dtic/tr/fulltext/u2/a199739.pdf

Pattillo, Donald M., *Pushing the Envelope: The American Aircraft Industry*, Ann Arbor, Mich.: University of Michigan Press, 2000.

Perry, Robert L., *Antecedents of the X-1*, Santa Monica, Calif.: RAND Corporation, P-3154, June 1965. As of February 29, 2012:
http://www.rand.org/pubs/papers/P3154.html

———, *Variable Sweep: A Case History of Multiple Re-Innovation*, Santa Monica, Calif.: RAND Corporation, P-3459, October 1966. As of February 29, 2012:
http://www.rand.org/pubs/papers/P3459.html

———, *A Prototype Strategy for Aircraft Development*, Santa Monica, Calif.: RAND Corporation, July 1972, not available to the general public.

Perry, Robert L., D. DiSalvo, George R. Hall, Alvin J. Harman, G. S. Levenson, Giles K. Smith, and James P. Stucker, *System Acquisition Experience*, Santa Monica, Calif.: RAND Corporation, RM-6072-PR, November 1969. As of February 29, 2012:
http://www.rand.org/pubs/research_memoranda/RM6072.html

Pierrot, Lane, *Options for Fighter and Attack Aircraft: Costs and Capabilities*, Washington, D.C.: Congressional Budget Office staff memorandum, May 1993.

Pierrot, Lane, and Jo Ann Vines, *A Look at Tomorrow's Tactical Air Forces*, Washington, D.C.: Congress of the United States, Congressional Budget Office study, January 1997.

Pint, Ellen M., and Rachel Schmidt, *Financial Condition of U.S. Military Aircraft Prime Contractors*, Santa Monica, Calif.: RAND Corporation, MR-372-AF, 1994. As of February 29, 2012:
http://www.rand.org/pubs/monograph_reports/MR372.html

Rich, Ben R., and Leo Janos, *Skunk Works: A Personal Memoir of My Years at Lockheed*, Boston, Mass.: Little, Brown, 1994.

Rich, Michael, and Edmund Dews, with C. L. Batten, Jr., *Improving the Military Acquisition Process: Lessons from RAND Research*, Santa Monica, Calif.: RAND Corporation, R-3373-AF/RC, 1986. As of February 29, 2012:
http://www.rand.org/pubs/reports/R3373.html

Scherer, F. M., "Market Structure and the Employment of Scientists and Engineers," *American Economic Review*, Vol. 57, No. 3, June 1967, pp. 524–531.

Schmutzler, Armin, *The Relation Between Competition and Innovation: Why Is It Such a Mess?* London: Centre for Economic Policy Research Discussion Paper 7640, January 2010.

Scutts, J. C., *F-105 Thunderchief*, London: Ian Allen, 1981.

Shenon, Philip, "Jet Makers Preparing Bids for a Rich Pentagon Prize," *New York Times*, March 12, 1996. As of April 9, 2012:
http://www.nytimes.com/1996/03/12/us/jet-makers-preparing-bids-for-a-rich-pentagon-prize.html

Smith, Giles K., and E. T. Friedmann, *An Analysis of Weapon System Acquisition Intervals, Past and Present*, Santa Monica, Calif.: RAND Corporation, R-2605-DR&E/AF, November 1980. As of February 29, 2012:
http://www.rand.org/pubs/reports/R2605.html

Smith, Giles K., A. A. Barbour, Thomas L. McNaugher, Michael D. Rich, and William Stanley, *The Use of Prototypes in Weapon System Development*, Santa Monica, Calif.: RAND Corporation, R-2345-AF, March 1981. As of February 29, 2012:
http://www.rand.org/pubs/reports/R2345.html

Sponsler, George C., N. Rubin, Dominique Gignoux, and E. Dare, *The F-4 and the F-14*, Gaithersburg, Md.: Columbia Research Corporation, May 1973.

Stevenson, James Perry, *Grumman F-14 "Tomcat,"* Fallbrook, Calif.: Aero Publishers, 1975.

———, *McDonnell Douglas F-15 Eagle*, Fallbrook, Calif.: Aero Publishers, 1978.

———, *The Pentagon Paradox: The Development of the F-18 Hornet*, Annapolis, Md.: Naval Institute Press, 1993.

———, *The $5 Billion Misunderstanding: The Collapse of the Navy's A-12 Stealth Bomber Program*, Annapolis, Md.: Naval Institute Press, 2001.

Stoff, Joshua, *The Thunder Factory: An Illustrated History of the Republic Aviation Corporation*, Osceola, Wis.: Motorbooks International, 1990.

Stuart, William G., *Northrop F-5 Case Study in Aircraft Design*, Hawthorne, Calif.: Northrop Corporation, September 1978.

Stubbing, Richard A., with Richard A. Mendel, *The Defense Game: An Insider Explores the Astonishing Realities of America's Defense Establishment*, New York: Harper and Row, 1986.

Sullivan, Michael, *Tactical Aircraft: F/A-22 and JSF Acquisition Plans and Implications for Tactical Aircraft Modernization*, Testimony Before the Subcommittee on AirLand, Committee on Armed Services, U.S. Senate, Washington, D.C.: U.S. Government Accountability Office, GAO-05-519T, April 6, 2005. As of June 27, 2012:
http://purl.access.gpo.gov/GPO/LPS60224

———, *Joint Strike Fighter: Impact of Recent Decisions on Program Risks*, Testimony Before the Subcommittees on Air and Land Forces, and Seapower and Expeditionary Forces, Committee on Armed Services, House of Representatives, Washington, D.C.: U.S. Government Accountability Office, GAO-08-569T, March 11, 2008. As of June 27, 2012:
http://www.gao.gov/new.items/d08569t.pdf

———, *Joint Strike Fighter: Strong Risk Management Essential as Program Enters Most Challenging Phase*, Testimony Before the Subcommittee on Air and Land Forces, Committee on Armed Services, House of Representatives, Washington, D.C.: U.S. Government Accountability Office, GAO-09-711T, May 20, 2009. As of June 27, 2012:
http://www.gao.gov/new.items/d09711t.pdf

———, *Joint Strike Fighter: Significant Challenges and Decisions Ahead*, Testimony before the Subcommittees on Air and Land Forces and Seapower and Expeditionary Forces, Committee on Armed Services, House of Representatives, Washington, D.C.: U.S. Government Accountability Office, GAO-10-478T, March 24, 2010. As of June 27, 2012:
http://purl.access.gpo.gov/GPO/LPS122323

———, *Joint Strike Fighter: Restructuring Places Program on Firmer Footing, but Progress Is Still Lagging*, Testimony Before the Committee on Armed Services, U.S. Senate, Washington, D.C.: U.S. Government Accountability Office, GAO-11-677T, May 19, 2011. As of June 27, 2012:
http://www.gao.gov/new.items/d11677t.pdf

Swanborough, Gordon, *United States Military Aircraft Since 1909*, London: Putnam, 1963.

———, *United States Navy Aircraft Since 1911*, London: Putnam, 1968.

Sweetman, Bill, *A-10 Thunderbolt II*, New York: Arco Publishing, 1984a.

———, *Phantom*, London: Jane's Publishing Company Limited, 1984b.

———, "Lockheed YF-22: Stealth with Agility," *World Airpower Journal*, Vol. 6, Summer 1991a.

———, "The Fighter They Didn't Want," *World Airpower Journal*, Vol. 7, Autumn–Winter 1991b.

Sweetman, Bill, and James C. Goodall, *Lockheed F-117A: Operation and Development of the Stealth Fighter*, Osceola, Wis.: Motorbooks International, 1990.

Thornborough, Anthony M., and Peter E. Davies, *Grumman A-6 Intruder Prowler*, London: Ian Allen, 1987.

U.S. Government Accountability Office, *Tactical Aircraft: Opportunity to Reduce Risks in the Joint Strike Fighter Program with Different Acquisition Strategy*, Washington, D.C., GAO-05-271, March 2005. As of February 29, 2012:
http://purl.access.gpo.gov/GPO/LPS59279

—————, *Joint Strike Fighter: DOD Plans to Enter Production before Testing Demonstrates Acceptable Performance*, Washington, D.C., GAO-06-356, March 2006. As of June 27, 2012:
http://purl.access.gpo.gov/GPO/LPS68153

—————, *Joint Strike Fighter: Progress Made and Challenges Remain*, Washington, D.C., GAO-07-360, March 2007. As of February 29, 2012:
http://purl.access.gpo.gov/GPO/LPS80611

—————, *Joint Strike Fighter: Recent Decisions by DoD Add to Program Risks*, Washington, D.C., GAO-08-388, March 2008. As of February 29, 2012:
http://www.gao.gov/new.items/d08388.pdf

—————, *Joint Strike Fighter: Additional Costs and Delays Risk Not Meeting Warfighter Requirements on Time*, Washington, D.C., GAO-10-382, March 2010. As of February 29, 2012:
http://www.gao.gov/new.items/d10382.pdf

U.S. House of Representatives, *The Navy's A-12 Aircraft Program: Joint Hearing Before the Procurement and Military Nuclear Systems Subcommittee and the Research and Development Subcommittee and the Investigations Subcommittee of the Committee on Armed Services, U.S. House of Representatives*, Washington, D.C., December 10, 1990.

—————, *A-12 Acquisition: Hearings Before the Investigations Subcommittee of the Committee on Armed Services, U.S. House of Representatives*, Washington, D.C., April 9, April 18, July 18, July 23, July 24, 1991a.

—————, *Oversight Hearing on the A-12 Navy Aircraft*: Hearings Before the Legislation and National Security Subcommittee of the Committee on Government Operations, U.S. House of Representatives, Washington, D.C., April 11, July 24, 1991b.

Yenne, Bill, *Lockheed*, New York: Crescent Books, 1980.

—————, *McDonnell Douglas: A Tale of Two Giants*, New York: Crescent Books, 1985.